Residential Land Development Practices

Other Titles of Interest

Engineering for Sustainable Human Development, by Bernard Amedei. (ASCE Press, 2014). Offers a framework and guidelines for the delivery of small-scale development projects that combine concepts and tools traditionally used by development agencies with techniques from engineering project management. (ISBN 978-0-7844-1353-1)

So Your Home is Built on Expansive Soils: A Discussion on How Expansive Soils Affect Buildings, edited by Warren K. Wray, Ph.D., P.E.; Marshall B. Addison, Ph.D., P.E.; and Kenneth M. Struzyk, P.E. (ASCE, 2019). Written for the layperson in clear, easy-to-understand terms to assist homeowners in understanding why some soils and conditions can lead to structural or cosmetic damage to buildings. (ISBN 978-0-7844-1521-4)

Construction Contract Claims, Changes, and Dispute Resolution, by Paul Levin (ASCE Press, 2016). Guides engineers and contractors through the complex process of construction contracting, focusing on claims and changes during the construction process. (ISBN 978-0-7844-1429-3)

Engineering for Sustainable Communities: Principles and Practices, edited by William E. Kelly, Ph.D., P.E; Barbara Luke, Ph.D., P.E.; and Richard N. Wright, Ph.D. (ASCE, 2017). Defines and outlines sustainable engineering methods for real-world engineering projects. (ISBN 978-0-7844-1481-1)

Economics and Finance for Engineers and Planners, by Neil S. Grigg, Ph.D., P.E. (ASCE Press, 2010). Presents the core issues of economics and finance that relate directly to the work of civil engineers, construction managers, and public works and utility officials. (ISBN 978-0-7844-0974-9)

Residential Land Development Practices

A Textbook on Developing Land into Finished Lots

Fourth Edition

David E. Johnson, P.E., F.ASCE

ASCE PRESS

Library of Congress Cataloging-in-Publication Data

Names: Johnson, David E., 1950 September 28- author. | American Society of Civil Engineers.
Title: Residential land development : a textbook on developing land into finished lots / David E. Johnson P.E.
Description: 4th Edition. | Reston : American Society of Civil Engineers, 2021. | Revised edition of the author's Residential land development practices, c2008. | Includes bibliographical references and index. | Summary: "David E. Johnson shows owners, local officials, and real estate professionals how to navigate the opportunities and pitfalls that encompass residential development projects"– Provided by publisher.
Identifiers: LCCN 2020040089 | ISBN 9780784415672 (paperback) | ISBN 9780784483275 (ebook) | ISBN 9780784483282 (epub)
Subjects: LCSH: Residential real estate. | Land subdivision. | Real estate development.
Classification: LCC HD1390.5 .J64 2021 | DDC 333.77–dc23
LC record available at https://lccn.loc.gov/2020040089

Published by American Society of Civil Engineers
1801 Alexander Bell Drive
Reston, Virginia 20191-4382
www.asce.org/bookstore | ascelibrary.org

***Errata:* Errata, if any, can be found at https://doi.org/10.1061/9780784415672.**

ISBN 978-0-7844-1567-2 (print)
ISBN 978-0-7844-8327-5 (PDF)
ISBN 978-0-7844-8328-2 (ePub)

Manufactured in the United States of America.

27 26 25 24 23 22 3 4 5 6

Dedicated to Lily and Emma

Disclaimer

This textbook on residential land development practices is based on the field and management experience of the author. All of the concepts presented in this book have been successfully used by builders and land developers in practical applications. This book is not intended to present all available land development management techniques. The figures contained within are a result of working with many professionals on land development projects. The sample contracts contained in this book are to be used as guides in preparing specific project agreements. Binding contract agreements should be prepared and/or reviewed by an attorney.

Contents

It is a great profession. There is the fascination of watching a figment of the imagination emerge through the aid of science to a plan on paper.

Then it moves to realization in stone or metal or energy.

Then it brings jobs and homes to men.

Then it elevates the standards of living and adds to the comforts of life.

That is the engineer's high privilege.

To the engineer falls the job of clothing the bare bones of science with life, comfort, and hope.

No doubt as years go by the people forget which engineer did it, even if they ever knew.

But the engineer himself looks back at the unending stream of goodness which flows from his successes with satisfactions that few professions may know.

—Herbert Hoover, 31st US president,
who was an engineer before becoming president

Introduction

This is a comprehensive textbook on how to develop vacant land into marketable residential lots and homes. The land development profession is learned through experience, not taught. Design professionals, developers, and builders use each project experience to their benefit during the development of subsequent residential subdivisions. Land development experts do not exist. It is the level of experience attained by each land development stakeholder that can influence the success of a project. Each local jurisdiction has its own process of regulating, debating, designing, and building residential communities. Because the industry is fragmented and locally managed, this book provides a generalized approach to developing property. The overall land development approach outlined in this book should be materially changed to reflect the nuances of each jurisdiction.

This book uses the tools of the trade to illustrate how to find land and develop the land into finished lots for single-family homes. Civil engineers, land development beginners, or seasoned professionals associated with building our residential neighborhoods and communities can use this book to learn and better understand the complexities of the residential land development business.

The unhindered entrepreneurial vision of the builder and developer in the pursuit of producing a quality and sustainable living environment must be preserved. We must learn how to balance all the vested interest groups involved in the land development process to meet the challenge of bringing raw land to finished lots and houses.

Target Readership

This textbook will benefit a wide spectrum of stakeholders and those entering the land development industry, including the following:

- The developer, who constructs finished lots from raw land;
- The builder, who constructs the houses on a subdivided and/or finished lot;
- The engineer, who designs the subdivisions and signs and seals the design plans;
- The architect, who prepares the house plans;
- Local, regional, state, and/or federal government staff, who review and approve the engineering construction documents for each project;
- The real estate agent, who ultimately sells a home on the improved lot;
- The home-buying public, who buys lots and/or homes;
- The public, who may want to influence the project outcomes; and
- The planning board, board of adjustment, or council members, who cast their votes on the merits of a subdivision plan and influence how a neighborhood is designed and constructed.

ASCE and the National Association of Home Builders (NAHB) are the primary organizations whose members are engaged in the overall process of converting raw land into finished lots.

Chapter 1

Land Planning Concepts

The residential building industry, over many years, has progressed in the use of land and in the development of planning concepts. Although many communities vary their zoning districts to change densities, land planning approaches have not changed dramatically over the years. The land planning concepts of many communities started with the pattern found in many older towns and cities. Neighborhoods in which homes were close to the roadway and close to neighbors formed a sense of community and a sense of belonging. Duplicating these concepts is difficult, especially with rigid bulk requirements imposed by local government. Many communities are leaning toward conditional zoning and leaving creativity and ingenuity to the designer rather than the regulators.

Creating a sense of community within a residential community can be achieved if the project is of sufficient size to effectively and financially create a lifestyle the market can embrace. The larger the parcel of land and the greater the number of housing units with a commercial/industrial component, the greater flexibility designers have in creating a land plan unique to the property and to the marketplace. However, not every community will embrace land planning approaches tried in other communities. Nor can developers always amass sufficient land holdings to create a new community concept. Also, many developers do not want to take the chance on a new land planning concept because of the risk of government and market rejection of the idea. Thus, local governments have instituted land planning approaches in the zoning ordinance of sufficient depth to guide development patterns. These regulatory design approaches legislate the way people should live without regard to how people want to live.

The market dictates the sales price of a home. The zoning dictates the number of homes, and the land dictates the success of the project. After each peak in a housing cycle, the building industry always faces efforts by local governments and residents to stifle continued residential growth. In fact, many jurisdictions downzone properties to achieve less dense projects. Local government should, however, be assessing land availability, existing infrastructure, and density factors to encourage infill or smart growth patterns rather than encourage suburbanization of a community.

Affordable Homes

When local governments institute regulations on land use that impede growth or financial profitability, the building community must seek out less expensive land to meet the demands of the marketplace. Restrictive planning approaches will significantly affect the way communities are developed for years to come. Exhausting land resources through low-density development and overextending community infrastructure servicing these developments will halt housing development in many parts of the country.

The affordability gap in the housing market continues to widen. The lack of affordable housing for low- and moderate-income home buyers is an acute problem for many communities. A concerted effort between regulators, builders, and engineers is necessary to bridge this housing affordability gap. To achieve affordability, builders need flexibility in the way land can be used and in the standards that regulate how the land is developed. The price of land does not correlate to the marketability or acceptability of the housing product. Land designated with higher densities allows builders to spread the cost of land, designs and permits, and land development costs over more units, thus permitting builders to reach the first-time home buying market and attain a higher level of sustainability.

Land Planning for Marketability

Within many housing markets, the competition for home buyers is always keen. A housing project distinguishes itself from its competition with the design of the home and the land planning approach. The design of the home can easily be offset by another builder with price adjustments or product enhancements, but the site characteristics and infrastructure planning cannot significantly be adjusted after approvals and permits have been granted. Significant tree-save areas within a subdivision, for example, can enhance the ambiance of the street scene.

Land planning for marketability is not a new concept. The creativity of the land plan, site amenities, street scene, aesthetic presentation, and environmental considerations are all ingredients of a successful housing project. These elements create the sense of community all buyers seek and set the project apart from the competition. A well-thought-out and visualized land plan with attractive street scenes combined with usable, livable, and properly oriented lots will increase absorption rates and the perceived value of the homes.

Engineers should be elevating their land planning and subdivision design skills to the next level of attention. A housing project need not be designed in a vacuum, and each housing lot should be considered as an individual project. Respecting the end user is essential for a successful project; and proper grading, orientation of the house, varying setbacks, landscaping, and lot usability are key elements. Engineers should pay special attention to the location of driveways and garages, and to the lot topography so as to maintain existing natural features of the lot. The less grading and clearing on a lot, the better the product is received by the home buying public.

Although there are valid reasons to incorporate different land planning concepts in the marketplace to present a clear choice, many new land planning concepts are not embraced by the public and the regulators. Most locales have their own approaches as to how land is developed, and bulk requirements vary from one location to the next. Many municipalities try to legislate how land can be used for housing developments. Unfortunately, this legislative approach is not market sensitive, and many projects do not move forward because builders are unwilling to raise sales prices because of increased regulatory costs. Residential land planning has everything to do with the desire of the marketplace. The density, style of home, and size of project become secondary to the land planning and engineering design of the project.

Land prices match the demand but very rarely depreciate. Thus, builders must raise house prices, achieve higher densities, or seek outlying areas with lower land prices. Because builders cannot continue to raise prices while government controls the density, builders must seek fringe or suburban areas to continue their building programs. This has caused a reaction from many municipalities and their citizens to slow or curb housing development. Knowing that land prices do not substantially depreciate in a slow economy, the pent-up home buying market will make concessions and purchase a home in outlying areas. This stresses infrastructure, adds to transportation costs, and causes government to downzone property. A better approach would be to find ways to concentrate growth in areas where the government considers housing a must and manage the existing infrastructure to match development patterns.

As regulators, builders, and designers search for cost-effective and consumer-driven projects, there are four land planning concepts that would provide regulatory compliance, achieve affordability, and be considered sustainable: high density, cluster, neotraditional, and active adult.

High-Density Developments

Most zoning approaches earmark high-density developments close to other nonresidential uses, existing infrastructure, and difficult sites. Land planning concepts for higher-density projects are governed by the housing type accepted by the marketplace. In many locales, zero lot line or small lot single-family subdivisions are not well received, but in many municipalities high-density single-family housing is the only answer to the complicated housing equation. High-density projects concentrate on amenities, such as views and open-space planning of the project. Flexibility in road or parking lot design further generates creative solutions to difficult land forms. Engineers should concentrate on how the housing units relate to each other and plan the infrastructure around the housing orientations, not the other way around. Higher-density developments such as zero lot line, small single-family lots; patio homes; duplexes; fourplexes; townhouses; and low-, mid-, and multistory buildings may be limited by some jurisdictions to control growth. This is exactly the wrong approach if sensible growth patterns and meeting the pent-up demand for affordable housing is a priority. Higher density will give builders a better opportunity to offer affordable housing prices to first-time home buyers.

Cluster Zoning

Cluster zoning is being implemented by many jurisdictions with the goal of increasing open space within a project. In essence, a cluster zoning deed restricts a portion of the project and consolidates the housing in one location (in general, on smaller lots) to achieve the density permitted within the gross acreage of the site. This approach has an obvious benefit to the taxpaying community by limiting the cost of infrastructure maintenance and community services. With the price of land, even marginal land is a problem for developers trying to make cluster projects financially successful. If smaller lot sizes and increased net densities are combined with cluster zoning practices, local government, developers, and home buyers benefit. Cluster zoning is a tool that can be successful in the correct circumstances; however, the land characteristics have a direct bearing on the success of a cluster land plan. Cluster zoning should not be restricted to a single zone within a municipality, but rather a flexible planning option should be permitted in all zones. Although this planning approach is not new, cluster provisions are not widely used because builders are hesitant to use them without greater assurance of the increased densities necessary to manage land costs.

Neotraditional

This land planning concept attempts to replicate the community design approaches of traditional urban planning by de-emphasizing garages and cars while emphasizing humanization. Today's neotraditional approach has elements and street scenes similar to older, well-established communities. This advanced pedestrian-friendly planning approach uses smaller streets, street trees, small lot frontages, close front setbacks, and garages tucked into the backyard to benefit the end user—the home buyer. The concept is valid and can be quite successful in the correct environment. Garages can connect to an alleyway behind the lots while the homes face the residential streets. Builders will attempt this land planning concept if the regulations governing the street and alleyway design are modified and if the standards governing the infrastructure design are made more flexible. The land development cost for neotraditional plans will be higher, but it can be offset by the higher density associated with this type of development. Even if the regulations are flexible and the density factors advantageous, builders will try this approach only if it provides a competitive edge in the marketplace.

Adult Retirement Housing

Creative land planning approaches to retirement community design are essential to a successful project. In many communities, the retirement market is only starting to emerge. Engineers must understand and appreciate the retirement market and those elements of the community design that are necessary for a project to succeed. Different levels of retirement community design are based on the age group and services required by that group. This market segment will only increase as baby boomers age. Serving this market segment with new housing and alternative housing concepts will generate new approaches to land planning and regulatory control. In addition, the existing housing stock will turn over, helping to bridge the affordability gap in many communities. If engineers have the land planning skills, expertise, and creativity for this market segment, their client base will expand their business by word of mouth.

Connectivity

Regardless of the land planning approach, most municipalities require *connectivity*, that is, the linkage of roads within a project and access to adjacent undeveloped property by the elimination of cul-de-sacs. Regulatory planners seek social interconnections within a housing project, because their goal is to enhance social good and re-create an urban environment. Engineers should accommodate connectivity as a practical convenience for neighboring properties. However, eliminating cul-de-sacs by providing connections between streets is contrary to the way most families want to live. Most families prefer to live on cul-de-sacs or short streets with limited traffic; not only are conflicts between vehicles and pedestrians minimized, but homeowners consider the design safer and more secure.

Partnerships in Planning

Housing professionals and the regulatory community should partner together in seeking design alternatives that are market driven, rather than governed solely by regulation. In some cases, increased cost of development and market rejection of a housing concept are the real unintended consequences of a failed planning concept based on a flawed vision.

State and local planning officials react to a strong housing market with smart growth initiatives that direct housing projects toward sites with existing infrastructure. The problem with this approach is that the cost of land is usually higher when infrastructure is already in place. Raw land as a commodity is the one factor that cannot be controlled. The need for someone to sell land is primarily an economic decision; builders will purchase land at a premium only if the market can absorb the higher housing prices. If the market demand is for a lower sales price, the builder will seek fringe, suburban, or infill properties so that higher densities can offset the cost of land. Land values do not always follow supply and demand curves or remain consistent with historical sales prices.

Another government reaction to overdevelopment is to downzone properties. As a result, land values increase on a per-unit basis, and home prices are driven higher. Again, the gap in affordable housing widens as the government reacts to housing growth.

A partnership among engineers, builders, and government officials is important in developing a truly smart growth initiative that properly builds on existing infrastructure and meets the demands of each housing segment. Every community has unique housing demands, and housing solutions should be a result of consensus among the various groups involved in the design, approval, and construction of homes. Engineers have an opportunity to be key participants in the development of housing, and they must take an active local role in the development of housing ideas, concepts, and programs.

Chapter 2

Residential Land Uses

For each parcel of land considered for purchase, builders must assess the surrounding land uses for compatibility with the proposed use. If the municipal land-use and zoning maps indicate a specific use (such as residential) for the property in question and it is compatible with surrounding land uses, rezoning may not be justified. Proceeding with a residential use consistent with a government's intended use for the property will reduce the regulatory process. Not all municipalities would grant the highest and best use for the property even if the property could be developed efficiently, be supported by adequate existing infrastructure, and be financially feasible for the developer and the community. But they would entertain a zoning change complementing the vicinity. In areas with agricultural land uses, any land development idea, including small lot residential, would necessitate a rezoning. The builder should use the market study and assess the market demands as a part of the due diligence process.

The market study will define housing design programs for the strongest market segment. The study should also investigate other market segments to determine a need for commercial retail, offices, industrial buildings, and other nonresidential uses. If the property is zoned residential but the market study illustrates strong market metrics for nonresidential uses, the purchaser of the property should consider rezoning opportunities.

The site analysis report should contain technical considerations outlining the advantages and disadvantages of developing the property in question. This report should also consider the land uses most advantageous to develop from a technical viewpoint. As an example, the site analysis report and market study will be used by the design team to create a land-use plan and land plan for a specific design program and property. If the market study indicates industrial uses are needed but the site analysis identifies steep slopes on the property with no public water available to the site, the property may not be suitable for a large industrial building. Residential land uses are the most common uses in most municipalities. Complementing the land form with housing style and price point will result in a project that is marketable, cost-effective, and sustainable for the community.

Because the land development business is a fragmented industry with different market considerations and price points across the country, selecting a land use suitable for the property characteristics, market, and location is essential for a successful project. This chapter will briefly describe land-use alternatives and options available in most municipalities across the country.

Rural

Distinguishing factors for rural residential use are large lots in excess of 1 acre (ac) with wide road frontages and lots typically located beyond public water and sewer services. The building setbacks are typically generous to correspond with the lot dimensions. The road infrastructure may be designed using different criteria because the traffic volumes are expected to be low. The lots may be served by roadways without the curbs and sidewalks usually found in higher-density residential projects. Nonresidential uses may not be suitable for rural areas, because residential uses usually precede the need for nonresidential uses. Some municipalities use rural residential lot standards to downzone property. In these areas, the minimum lot size may be 3 to 5 ac and lot frontages may increase to a minimum of 200 ft. Without corresponding land development standards and criteria, developing a rural residential area may be cost prohibitive. The infrastructure design standards for this use are important for managing the project costs.

Estate

Estate residential classification includes various lot sizes but is set apart by the housing product, price point, and site features (Figure 2-1). The lot sizes may vary from one-half to 1 acre, but the lot frontage may vary from 90 to 100 ft. Because the homes may be customized, they are typically wider and may include three-car garages. With high rooflines, varying architectural material, and wide front façades, the street scene will be accentuated. Estate residential areas are situated in prime locations or associated with unique site features. Estate housing may be located around golf courses, planned for secluded locations, or arranged for

Figure 2-1. Estate residential use. Single-family homes with the highest price points in the community are situated on estate-sized lots with wide frontages.
Source: Sidney T. Burns, Due Diligence Associates, Parkland, Florida.

advantageous views or vistas. Site characteristics and uniqueness complement the price points of higher-end housing.

A gated community may necessitate private roads, which may be designed at different standards than public roads. The roadway could be contained in an ingress–egress easement rather than a road right-of-way. The additional land not incorporated in the right-of-way is included in the lots. Road design should incorporate traffic-calming design approaches to reduce vehicle speed within the community. Hardscape and softscape features are an important design element for road frontage. Substantial landscaping and security features provide perceived value and a sense of exclusivity. Most master-planned large-scale communities will have an estate residential area with slower absorption rates than other market segments. The estate area will also have more stringent covenants, conditions, and restrictions designed to protect property values.

Low Density

Presently, most residential subdivisions are constructed as low density to satisfy the move-up market (Figure 2-2). This market segment has the greatest depth and is probably the most sensitive to interest rate fluctuations. Price points vary significantly from one market to another. Low-density residential areas contain two to four units per acre. Lot areas will vary from 10,000 to 20,000 ft^2. In general, the front and rear yard setbacks are 25 ft, and side yard setbacks are 10 ft. Lot width varies from 85 to 100 ft, which permits a variety of architectural designs for conventional housing.

The roadway design is also standard in terms of street width, curbs, sidewalks, and tree plantings. Yard swales in lieu of curbing may return to the marketplace as a result of recent emphasis on best management practices. The marketability of this type of project, however, may be

Figure 2-2. Low-density residential use. Many projects located beyond public services will use on-site water and wastewater treatment systems requiring larger lot sizes. The ability to save trees, reduce pavement widths, and eliminate curbs are consistent design approaches for rural single-family subdivisions.
Source: David E. Johnson.

difficult because of the historical design approach for conventional subdivisions.

Small Lot, Single Family

Higher-density, detached, single-family housing has become the use of choice for many builders. Land prices continue to escalate but the cost may be spread over more units and the buyer offered a detached house. In slow markets, prices for homes soften but the price of land stabilizes. In most cases, the percent decrease in home prices will not be reflected in the price for vacant land. Small lot, single-family land use provides the most advantageous scenario for a developer, who can spread the cost of land over more housing units and also provide a product that the market is seeking (a detached single-family house). In many markets, the trend is to provide small, single-family lots but larger houses (Figure 2-3).

A variation on small lot, single-family homes are *zero lot line* or *patio* homes. Zero lot line planning concepts were first introduced in the late 1970s but did not receive wide market acceptance until later. The lot width runs from 50 to 85 ft, and the lot depth runs from 90 to 120 ft. The house is positioned on a side property line, and the property line side of the house is designed with no openings or with opaque glass. The house design and accessibility are oriented toward the wider portion of the lot. In many projects, the side yards are fully landscaped and hardscape entrances are incorporated into the house design.

The house designs for the small lot, single-family subdivision should minimize the prominence of the garage. The 50-ft-wide lot will have a product width between 35 and 40 ft, so in some cases the front door is located on the side of the unit. The challenge is to preplan the street scene elements and establish architectural elements to attract buyers. The primary architectural element to be considered is the roofline, including the pitch and truss layout. Road infrastructure for these sites will not differ from the subdivision road standards for low-density residential lots. However, more driveways and yards will impede the design of proper lot grading and drainage. Side yard swales typically split the lot to convey runoff to the front and back of the lot. Placing garages side by side on adjacent lots allows flexibility in drainage design and reduces interference with lot improvements. In many markets, builders combine the small lot, single-family approach with neotraditional land planning, in which garages are accessed from alleyways along the back of the lots.

Figure 2-3. Small lot, single-family residential use. Narrow lot widths and larger house sizes have been accepted design practices for many communities. This design approach increases density and allows buffers and common open space distribution throughout the project. *Source:* David E. Johnson.

Townhouses

As density increases, site planning approaches will be dictated by the architectural layout that accommodates the greater density. The standard townhouse (Figure 2-4) is one house within a row of attached homes, but townhouses can also be arranged in a neotraditional layout. The minimum number of townhouses in a building cluster is two, also known as a *duplex* unit. In general, townhouses are marketed to the first-time home buyer or workforce market. One-story townhouses are an alternative for the active adult market. One-story townhouses may not have garages, and thus the site plan reflects parking lot layouts or driveways for garage units. Parking lots increase the impervious surface of the site. Stormwater management and best management practices facilities would

Figure 2-4. Townhouse residential use. Attached single-family homes on fee-simple lots are an acceptable product choice for all market segments. Architectural appeal and on-street parking with rear-loaded garages have been well received in many marketplaces. *Source:* David E. Johnson.

become an important design consideration, requiring land dedication for these types of facilities.

Multifamily builders should design townhouse projects in phases, thus limiting the infrastructure costs to correspond with the absorption rate. The building clusters should limit the number of building configurations and the number of units in each building. For improving design flexibility and managing construction cash flow, townhouse clusters usually contain three to six units per building. The important aspect of this approach is that developers can start a building with a single sale but limit the number of speculative units built along with the sold unit.

Townhouse building clusters should have offset front elevations and varying roof lines for architectural appeal. Combined with creative use of materials and site planning, the clusters will have individuality and streetscape appeal. Townhouse developments are usually constructed in the range of 8 to 12 units per acre. These projects also usually include an amenity complex appropriate for the target market. Additional land area can be designated for open space or recreational use as the density of the community increases. Townhouse projects are often located in areas associated with other high-density uses or as a residential buffer between single-family, detached houses and nonresidential uses. Townhouse projects can also provide a sense of community and belonging that must be reflected in the overall land plan and site amenities.

High Density

High-density housing products (Figure 2-5) are designed vertically; their architecture must be cost-effective and have streetscape appeal. Densities ranging from 12 to 16 units per acre can be achieved with two-story, walk-up units. These units can be sold as condominiums or leased as rental units. The building footprint dictates the site plan and infrastructure design for the project. In most instances, two-story, walk-up buildings do not have underground or underneath parking, but rather parking lots and access lanes. In general, the parking lots provide 9 by 18 ft parking spaces and 24-ft-wide access lanes. In some municipalities, parking stalls are reduced in size for small vehicles. The number of spaces per unit certainly varies, and the total number of spaces may be increased for visitor and amenity site parking.

Most high-density residential projects have an amenity complex designed with a pool and playground equipment for young families. The competition in this market range is strong, and the site amenities and architecture set the project apart from the competition. High-density residential projects require substantial impervious surface improvements. The drainage, stormwater management, and best management practice design is the most important design element, since upland area may be needed for drainage facilities rather than buildings. However, most municipal ordinances set a minimum for open space, which may include such drainage facilities. Pedestrian access within the project complex should take account of centralized postal facilities. Other design considerations include irrigation, entrance features, solid waste collection, recycling facilities, streetlights, landscaping, building separation, building orientation, curb type, security, and connectivity to nonresidential uses.

Figure 2-5. High-density residences. Condominiums or rental buildings in high-density zoned areas typically achieve a density of 12 to 16 units per acre. For high-density projects, engineers must consider circulation, parking, mail delivery, dumpster locations, lighting, and drainage facility design.
Source: David E. Johnson.

High Density, Multistory

The high-density, multistory residential building (Figure 2-6) picks up where the two-story walk-up leaves off. The multistory residential building starts with 16 units per acre and increases as the location, market, and municipality permits. Most multistory residential buildings are located close to public facilities and mass transit links. Multistory buildings can satisfy all market segments, including active adult, simply because these buildings are serviced by elevators. Parking facilities are underneath the building, in parking lots, or in adjacent parking structures.

The biggest challenge in infrastructure design of multistory residences is the interface with existing facilities, since most high-rise buildings

Figure 2-6. High-density, multistory residential use. A housing product achieving more than 16 units per acre will be a multistory building. Project planning of a high-density, multistory project provides opportunities to separate pedestrian traffic from vehicular movement. *Source:* David E. Johnson.

are located within developed urban areas and the sites are relatively small in relationship to the number of units. As with high-valued commercial property, facilities for stormwater management and best management practice approaches may be underground so as not to use valuable upland area, thus fully utilizing the site.

Most residential builders and developers do not undertake high-rise, high-density residential projects because, once the building is started, the entire building must be constructed before occupancy can begin, and many units are not sold when construction begins. For this type of housing, the cash flow requirements are substantial, regulatory approvals time-consuming, and delivery schedules lengthy.

Active Adult Communities

Active adults from 55 to 65 years old often start to sell their existing homes and move to a community structured and designed for them (Figure 2-7). The potential of the active adult market is staggering, and projects catering to the active adult market should not be overlooked for any location. The essence of the active adult and aging markets is to satisfy a local need, because most of those 55 and older desire to live close to family or their current homes.

Florida and Arizona will always be retirement meccas, but developers should consider local markets for opportunities to satisfy this expanding age group. In most municipalities, ordinances do not have specific zones permitting active adult communities; planned unit development ordinances may provide sufficient design flexibility for this type of project. If not, builders should pursue regulatory changes permitting design characteristics catering to the retirement market. In areas with

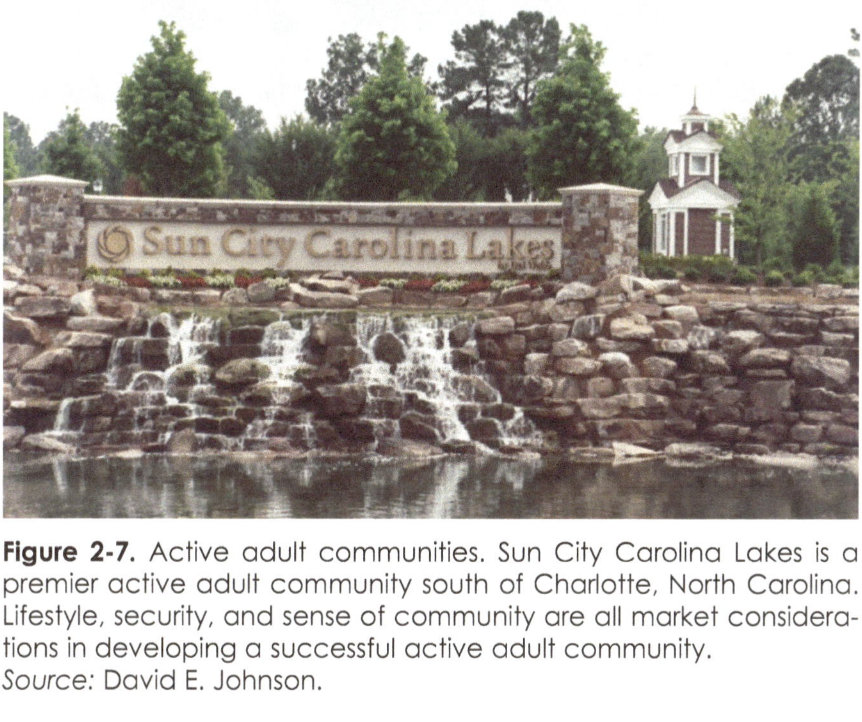

Figure 2-7. Active adult communities. Sun City Carolina Lakes is a premier active adult community south of Charlotte, North Carolina. Lifestyle, security, and sense of community are all market considerations in developing a successful active adult community.
Source: David E. Johnson.

active adult community zoning, eight units per acre is a density widely used by municipalities.

This specialty market requires attention to housing product lines and amenity packages because prospective homeowners are willing to spend money on upgrades and options in their new homes and they seek extensive recreation amenities and social activities. Most adult communities are structured with a homeowners' association that sets standards for maintaining individual yards and even the exteriors of homes. The monthly maintenance fee, if too high, is a marketing negative. Lot size is typically 50 ft wide and 90 ft deep.

The active adult market appears to prefer single-family, detached housing, but attached units can be successfully marketed to them. The attached units could range from duplexes to four or five units per building. With this building cluster approach, outside units are maximized and interior units limited. The end units are premium units, and the architectural design becomes more flexible with the additional exterior wall. Storage space must be considered in the design, and two-car garages are common. Multistory buildings are also viable options in the appropriate location.

Most active adult communities are designed with an amenity building catering to the social activities and recreation needs of this market. Golf courses, outdoor pools, walking tracks, exercise rooms, indoor pools, and pool tables are common features. Many communities have social activity directors on staff to manage the affairs of the recreation complex. The amenity buildings range in size but can be 30,000 ft^2 or larger. The amenity must be included in the initial design phases of the active adult community. Although there is no rule of thumb for the size of the project in terms of units, builders should seek a project size sufficient to carry the costs associated with the amenity package and other site features.

For the aging population, builders may consider a transitional community project, which includes the fee ownership of a single-family home or condominium and other levels of care options included with the same parcel of land. Care options include independent living, assisted living, congregate care, or continuing care facilities. The transitional community may also include multifamily attached units for alternative living arrangements and price points for expanding the market. A residential builder could consider a transitional community design when other care options can be marketed and constructed by an entity with expertise in those specific types of housing alternatives.

Traditional Neighborhood Developments

Traditional neighborhood developments provide a mix of housing opportunities and price points (Figure 2-8). Design elements include interconnection of roadway links to adjacent properties and existing neighborhoods or nonresidential uses. Each project has a recognizable core in the form of a balanced mix of nonresidential uses with pedestrian connectivity throughout the project limits. Within the project, neighborhoods are designed as neotraditional projects, which are also referred to as *new urbanism* (see Chapter 1). In most traditional neighborhood designs, cul-de-sacs are not preferred, and a hierarchy of interconnected roads guide the land planning approach.

Figure 2-8. Traditional neighborhood developments. Most traditional neighborhood development projects include neotraditional site design that locates garage access via backyard alleyways. These alleys are designed for one-way access and may be 14 to 18 ft in width.
Source: David E. Johnson.

As a current design trend, builders are recognizing the benefits of a well-planned community development that replicates small suburban town planning concepts in an urban setting. The emphasis on pedestrian-oriented streets is supported by instituting traffic-calming devices, such as roundabouts or narrow streets in the road network. Location of pocket parks, open-space elements, and recreation amenities also add to the overall development scheme. The attention to site detail and architectural design of the residential and nonresidential uses is the most important characteristic of a traditional neighborhood project. Land planning and infrastructure design establishes the framework, but the architecture will create the aesthetic appeal of the development, as well as street scenes marketable to a wide range of buyers.

Residential Mixed-Use Developments

Rezoning residential land into a mixed-use development is an effective way to increase the value of vacant land because it adds additional uses to a residential zone. A logical and successful land-use relationship is mixing residential uses with retail or business uses (Figure 2-9).

The residential use is the primary use, and it is integrated within the same building structure as the retail or business use. Many downtowns were originally planned with retail or business uses on the ground floor and residential units on the second floor. This arrangement of uses continues to be marketable and profitable for builders. This approach also widens the market, but the builder must have expertise in both commercial and residential construction, sales, and marketing. The architecture is extremely important to the marketability of mixed-use developments. The site plan layout is also important because the residential units will be used at different times of day than the retail or business units. Public versus private accessibility to the buildings and units must also be a consideration in the site layout. Vehicle movement and parking should be segregated, as well as access to each specific use. In many jurisdictions, the buildings are moved closer to the frontage road right-of-way. The buildings have two front elevations, with the building shielding the parking facilities from the frontage road. An alternative would be mixed-use zoning, which allows various uses within the same project boundaries but not within the same building. Many master plans or planned unit developments are mixed-use zoned properties.

Figure 2-9. Mixed-use developments. Commercial retail and residential housing components in a town center environment have proven successful and are considered a sustainable form of development. *Source:* David E. Johnson.

Large-Scale Projects

As municipalities continue to assess the pressures of growth and enact ordinances to guide future development patterns, builders must review their options of delivering profitable projects. For many years, land located in fringe areas was priced below the retail market because these areas lacked public infrastructure. Builders gravitated toward these fringe areas when lower land prices combined with a high market demand for housing beyond the urban core. Builders were willing to construct the infrastructure needed to serve these projects, but the results increased sprawl. Because the residential industry is fragmented, municipalities react to sprawl by revising ordinances so they discourage disjointed development patterns. Efforts for a consensus approach to development are overridden by public sentiment to enforce slow growth through government regulatory oversight and downzoning. Although this situation may seem detrimental to developing cost-effective residential projects, builders have an opportunity to present well-conceived projects based on sustainability and fostering positive reaction from local officials. The most common approach is preparing a planned unit development project based on defined goals, specific conditions, regulations, and standards.

In a well-conceived project, builders should devise a master plan with multiple parcels to illustrate how the land-use puzzle fits together with existing infrastructure. Such large-scale projects involve land controlled by the builder. The goal is not to spot-zone individual properties, but to assess an overall area to achieve a land-use plan that meets sustainability goals. Government and the land development industry must work together to plan defined neighborhood areas, an approach that should result in a consensus plan rather than a debate of the merits of another disjointed residential project.

Large-scale projects are beneficial for long-term growth and increased values. The property is approved as a whole, with approved conditions imposed by the developer and government agencies. This approval would vest the development rights for the property. The developer may seek changes as the market changes, but having a large-scale prospect with regulatory approvals reduces risk, saves time, and enhances the value of the property.

Chapter 3

Land Development Basics

The primary role of the land developer is to continuously maintain an inventory of buildable lots for housing construction. Not unlike any other industry, buying land and developing a product for the marketplace requires good business judgment and a keen sense of producing the product below the price points of competitors.

Marketable lots are a result of

1. Determining how the property will be used;
2. Completing a market study to determine the needs of the builder or the buying public;
3. Checking on comparable land values to ensure that the price is consistent with other subdivisions;
4. Evaluating the physical conditions of the land and its potential for subdividing;
5. Researching local, state, regional, and federal regulations affecting the site's development potential;
6. Understanding the legal issues associated with the land;
7. Knowing the cost of development;
8. Deciding to purchase of the land; and
9. Managing the land development process successfully.

Property Uses

A successful land development project is judged by knowing when lots will be released for sale, knowing how quickly the lots can be absorbed by the marketplace, and delivering the project on time. First, however, determine how the property will be used:

1. Include unimproved raw land in inventory without designs or approvals;
2. Sell unimproved raw land with approved designs and permits;
3. Sell approved lots and improved right-of-way infrastructure; or
4. Sell or use approved lots, improved right-of-way, and improved buildable lots.

These objectives should be evaluated in preparing business strategies and obtaining capital funding commitments. Pursuing each option allows a production-oriented business plan because the timing of the sale or use of lots under each scenario would be staggered. It is important to maintain an even production of lots. At the start of each project, a market study of location, housing products, competition, and land values must be completed based on tangible and valid research.

Market Study

Finding land to meet your business objectives becomes easier once the specific use of the land has been determined. Concentrate on the attributes of the overall community. Evaluate each specific neighborhood for development potential and evaluate each site within the neighborhood for its highest and best use.

Community Overview

Prepare a community overview by compiling and evaluating available information on employment, income, needs, and other demographic characteristics.

Demographic Analysis

Study the number of households, age distributions, population changes, and population trends.

Housing Factors

Study the housing inventory for owner-occupied units, rental units and vacancy ratios, trends in residential building activity, potential demand for the number of units expected to be built, and the annual construction rate of housing units in the community.

Economic Analysis

Study the employment trends, family income distribution, future income distribution, and unemployment rates.

Market survey firms, local chambers of commerce, regional planning boards, home builder associations, and other associated real estate agent groups compile specific economic data on communities, such as employment, population, demographics, and potential income by population segments. Economic surveys are complemented by evaluating the marketplace, talking with real estate agents familiar with school boundaries, and speaking with local residents. Read local newspapers for several months while studying the community; concentrate on the classified, business, and community sections of the newspaper. Drive through the community and identify competition by creating a map that shows the types of product, prices, and when the models are open.

Community Analysis

Visit economic development departments, building departments, zoning offices, chambers of commerce, and other sources to understand where people are moving from; family size; why they are moving to the community; and what nonresidential growth is expected for job generation. This information must be balanced with research on where people relax, their income levels, the number and distribution of families according to their household income, and other social factors. As a result of this research, focus on answering the following questions:

1. Can this community sustain the present building activity?
2. Can increased building activity be absorbed by the community?
3. What is the economic future and income expectancy for the area?
4. What are the future demographic characteristics of the area?

The result of this work will define competitive residential projects by housing type, price, size, features, and absorption rate for each model type.

Neighborhood Overview

After you have a complete analysis of the community, micro-analyze each selected neighborhood within the community. Evaluate the competition in the selected neighborhoods and ascertain whether business objectives can be achieved there. The true real estate axiom is *location, location, location*. However, evaluate site locations with a critical eye toward any adversarial neighborhood group that could react negatively to a new housing project. To further evaluate neighborhood potential, study local, county, and state new-home magazines. The product, land plan, and zoning requirements should be consistent with surrounding property. Evaluate the neighborhood's location, school boundaries, and geographic features.

Summary of Residential Market Analysis

- Determine housing needs and land development objectives.
- Document existing economic conditions within the area.
- Forecast the economic conditions of the area.
- Undertake a market study to analyze the housing market.
- Complete a competitive housing study of all current residential projects.
- Complete an affordability analysis comparing income with housing price.
- Summarize the data for sites under consideration.
- Recommend housing sales prices and raw land purchase goals.

Market Comparison

The proposed housing product line is selected to meet market demands and reflect the business goals of the project. The selection of competitive price points provides the framework to develop a marketable housing product. The selected product line is compared against the competition by using Appendix 3-1, a competitive housing market profile. These factors provide the microanalysis necessary in preparing a housing design program for the project.

Once you have selected several competitive projects for comparison, organize the research information in a resource file for ease of reference. Appendix 3-2 provides a list of site evaluation factors and the source from which this information can be obtained. Appendix 3-3 is a land evaluation scorebook that factors the attributes of the community and of each individual site. Using the scorebook approach will allow the evaluator to objectively outline the advantages and disadvantages of each site.

Site Analysis

Evaluating the site constraints of raw land is essential so as to minimize the financial risk involved in developing the land for housing. To minimize that risk, determine the lot yield and the cost required to properly develop the housing project before the purchase is executed. To do this, first undertake a complete comprehensive site analysis study (Figure 3-1). By combining this information with other elements outlined in this chapter, you can reduce the risk factors associated with each parcel under consideration.

Figure 3-1. Site analysis. A comprehensive site investigation and analysis of a site will identify development benefits and constraints. Electric transmission line rights-of-way will present development and marketing difficulties for a project.
Source: David E. Johnson.

No two parcels or sites have the same constraints or development solutions. It is imperative to understand those constraints and solutions so that houses can be delivered on time, absorbed rapidly by the marketplace, and constructed within budget. The more information that is known about a site and its physical characteristics, the better the chance of also knowing the cost implications, and the ultimate success or failure of the project can be determined.

Based on the results of the in-depth site analysis, determine the lot yield, prepare the conceptual land plan, understand the costs associated with the land plan, and select the housing product. During the investigative process, strive to identify the unknowns and determine why this property has not been purchased by competitors. As the unknowns are discovered, satisfied, and eliminated, the question becomes, "Why does this parcel remain available for sale?" and the risk of developing that land is further minimized.

The next step is to allocate financial resources for further site analysis and investigative work. Then, rather than focus on the reasons to purchase the property, establish all of the reasons *not* to purchase the property. This negative approach is beneficial during the regulatory process and helps to prepare a better product.

Site Analysis and the Purchase Agreement

Home building success is based on many factors. Among the most important are the value and cost relationship of raw land and the cost of developing that land for housing products. Ultimately, the results of the site analysis will be used to prepare the legal documents for the purchase of the land. The purchase agreement should use the identified site constraints as cause to void the contract without financial loss. For instance, if you are purchasing residential land that uses septic systems, then the soil test results must be favorable for septic systems. If the soil test results are not favorable, this would be sufficient cause to terminate the contract. The contract should also be contingent on receiving septic permits for construction before closing.

Once a contract has been prepared for the purchase of property, including a due-diligence clause, start the comprehensive site analysis work without delay. Obtain the information outlined in Appendix 3-2 before visiting the specific sites under consideration. Determining the physical and regulatory parameters governing each parcel gives a clear indication of whether the development of that parcel will be commercially successful. Appendix 3-4 is a detailed site analysis checklist. If this checklist is used to evaluate land, the physical investigation of the site can be clearly documented, illustrated on a plan for further clarification, and used as the basis for purchase decisions. This documentation can be valuable in obtaining financial commitments from lending institutions and in discussions with regulatory agencies or political groups.

Site Inspection

Once you have reviewed the information, you should walk the property and assess whether the information you have on the property's topography is accurate. Relate the elevation of the site to the elevation of the existing roads bordering the parcels. For example, if the property falls away from the road, the design approach will be different than if the property is uphill from the road. Moreover, the relative position of the property to its frontage roads has a bearing on the land development costs. The due-diligence clause of a purchase agreement should always require a soil boring or soil test pit program. Soil characteristics affect the site design and have a direct bearing on the cost of developing the land:

- If there is rock, basements are much more difficult to construct, and underground utilities are more costly to install.
- If there is clay material and if septic systems are planned, the clay material may have to be removed and replaced with a different material at significant cost.
- If the site has expansive soils, structural foundation designs for residential housing would be required; the cost of housing and associated liabilities would be much greater.

By walking the property, you should be able to confirm that the information obtained for the site is usually reliable. If a boundary and topography survey exists before the site inspection, the potential cut-and-fill problems on the site can be documented before design consultants are hired. The soil survey information is necessary to determine the seasonal high-water table elevation on the property, which is important for basement construction, installation of septic systems, dewatering for water and sewer systems, and locations of unmapped wetland areas.

Locating floodplains, freshwater wetlands, and natural drainage swales on the property will guide the drainage design for the project and affect the amount of buildable area. Thus, the profitability of the project can be severely eroded if these environmental constraints are not addressed during purchase negotiations.

Depending on the location of the natural drainage areas, the land plan of the project can be affected, and in some cases, the property can be rendered effectively useless. Most landowners do not understand how significantly governmental regulations can affect the marketability of the land. If a floodplain on a property has been documented by FEMA, the buyer of that lot will be required to obtain flood insurance. If the property is in the floodplain, the property may be approved for a no-rise permit. This would allow filling of the floodplain for development purposes. The *floodplain* is the primary conveyance of floodwaters and it includes the main channel and adjacent overbank areas. The *floodway* is the natural stream channel that conveys floodwaters. The floodway is typically not infringed on by development.

If the property is wooded, the vegetation on the property should be identified and located. If the trees are significant in size and type, they are excellent selling tools and positive attributes. If it is important to save trees for marketing purposes, mark on a field survey where the significant specimen trees are located. This information is essential to the land plan of the subdivision and affects the cost of clearing. If clear-cutting is designated on the land plan, the clearing cost per acre will be less, but the aesthetic appearance will suffer. If selective clearing is required to save special trees adjacent to rights-of-way, house envelopes, drainage structures, or utility facilities, the cost for clearing will substantially increase. The benefit of saving specimen trees, however, outweighs the cost associated with selective clearing.

When in the field, note on the boundary survey map the locations of public water distribution lines, fire hydrants, sanitary sewer manholes, electric service, telephone service, and cable television facilities. Indicate the location of each telephone pole along the roadway. These poles need to be field located to prevent conflicts with access drives, road widening, or utility installations. The cost of relocating utility poles can be extremely high. Make note of the frontage roadway width and quality of the pavement. Locate the nearest cross drains in the frontage street to determine whether off-site improvement is required.

Special Considerations

The following special characteristics must be noted during the site visit:

- Is there heavy truck or commuter traffic on the frontage road?
- Is there noticeable commercial or military air traffic?
- Is there unusual noise generated from nearby manufacturing plants or tire noise generated from interstate highway traffic?
- Are any unusual odors present? Note whether the odors would pose a negative marketing aspect.
- Are the odors from manufacturing plants or from agricultural uses?
- Are there any standing pockets of water? These may indicate wetlands or poor soils.

Take plenty of photographs of the site. Mark on a map the angle at which the pictures were taken. Photograph and describe any unsightly views. Also, photo document any views essential for the successful marketing of the project, such as a golf course, an ocean view, or other scenes.

Make note of any demolition requirements and junk or debris removal. Be certain to look for environmental concerns on the property, such as 55-gallon drums.

Make a clear notation of the compass orientation of the property in relationship to the surrounding land and views. Relate this information to the land plan and be sensitive to the way the houses would be oriented on the site.

During the site visit, walk along the exterior boundary of the property by pacing off distances in the absence of a survey, stakes, or flags. Take photographs of the surrounding area on all sides of the project. This photo history will be useful throughout the planning process.

Find the location of available natural gas lines and determine whether the lines can be extended to the site. Check for high-tension power lines, determine whether the electric company has future plans to install them, and learn whether easements exist for this type of improvement. Obtain deeds, tax maps, and other information to identify rights-of-way, easements, or restrictions on the site or on nearby properties.

If public water lines are available, check for sufficient pressure to ensure proper fire service. If pressure is lacking, you may be required to resize the lines, loop water lines, or provide water storage facilities, thus increasing your costs. If individual wells are planned, find out how deep the wells need to be drilled and whether the groundwater requires treatment. Discussions with well drillers are important in understanding any problems with individual wells; moreover, their knowledge of the site soil characteristics will be helpful in analyzing the site.

In the site analysis report, assign costs to the various factors that affect the development of the property. This cost analysis is essential in comparing the sites under consideration. For site access, determine if there are proper vertical and horizontal site distances at the expected access intersection at the site.

Also note whether the frontage road has vertical or horizontal curves or existing vegetation obstructing the line of sight at the access intersection. Determine the future plans of utility companies for extending facilities to the site and the time frame for completion. Determine whether the existing utility services will be adequate for the approval and permitting process. Determine whether a scheduling problem may result and include this item as a due-diligence condition of the purchase agreement.

Locate where the public sewer connection would be and determine whether the lay of the land from the street allows shallow gravity sewers. A low area within the property may require a pump station rather than balancing the cuts and fills to achieve a gravity sewer. Note this on the site analysis report.

The site analysis report should be comprehensive in order to make a sound business decision on the purchase of a site for housing development. Site analysis not only identifies cost elements but also illustrates potential positive or negative marketing aspects. The findings of an in-depth site analysis are the foundation for quality design, efficient planning, cost-effective engineering, and governmental acceptance.

Regulatory Issues

The development of land for housing is a regulated business and the cost is passed to the end user. The construction of a home within a subdivision is regulated and governed by the model building codes, whereas land and its use are governed by federal, state, regional, and local regulations, as well as self-imposed covenants and restrictions. These regulations are intended to protect the health, safety, and well-being of the citizens (Figure 3-2). Regulations affect the quality of our living environment and are intended to reduce the strains that new housing may place on municipal services.

Figure 3-2. Regulatory issues. Road typical sections are imposed at the local level. The road design criteria and right-of-way widths will dictate lot layouts. Residential roads without curbs and with narrow pavement widths provide an excellent best management practice design solution. Source: David E. Johnson.

Excessive Regulations

Regulations imposed by all levels of government result in delays engendered by the need to obtain approvals. The additional cost of regulatory compliance and approval delays are experienced by all builder/developers. Very rarely do we see government regulations rescinded, and they will certainly proliferate at all levels of government. Escalating land costs because of regulations increase housing costs, eventually resulting in a cost spiral that will exclude some people from the housing marketplace.

Most builders and developers must borrow capital before initiating a housing project, regardless of its size. The soft costs associated with preparing design plans, filing applications, and posting review escrows can be overwhelming. The longer it takes to obtain approvals, the more interest payments will mount. Sustaining overhead costs, monitoring labor costs, and managing material costs must be adequately addressed in the project pro forma by projecting how long it will take to obtain all permits and approvals. Forecasting cash flow needs that meet the requirements of the financial institution is critical for a builder or developer to hold onto the land during the approval process. In most states, the model building codes are adopted statewide and not amended at the local level; builders are aware of building code provisions and the costs associated with compliance. However, in land development, not only are design criteria, regulations, and standards imposed at the local level, but they can be duplicated regionally and/or by state agencies.

There can be inconsistencies in land development regulations between neighboring municipalities. The interpretation and inconsistencies cause unfair competition between developers and affect the marketability of finished lots. In most municipalities, a developer must post escrow money up front to pay municipal consultants or independent experts to review the project engineering designs for compliance with municipal regulations. The reviewing agent often interjects his or her own professional judgment and opinions (thus going beyond the intent of the regulations) into the design. This type of technical conflict delays the approval of the residential construction plans. Therefore, when consultants are selected, it is imperative to confirm that they understand the local political and regulatory climate. If the improvement plans can be prepared to meet the requirements, written or unwritten, and approved at the first hearing, the cost of doing business is minimized.

Any regulatory review, beyond the initial review, adds time to the approval process and can possibly cause a failure in meeting the closing goals of the company. During the governmental approval process, time is money and too much time may mean the potential loss of business. The goal of expediting the permitting of a well-designed residential project should be shared by government, private developers, and land development engineers.

Approval Process

Appendix 3-5 describes a typical residential construction approval process, illustrating the various steps within all levels of government chartered to process subdivision plans. This flowchart varies from state to state, but the basic technical requirements for subdivision approvals remain constant. The timeline has been adjusted to show ranges because of regulatory and regional differences. The only way to compress the approval process is to work with the design consultants and have them embrace the importance of the project goals. When negotiating land purchases, ensure that the purchase agreement addresses the regulatory process and the time needed to obtain permits and approvals before construction. Only consultants can compress the regulatory timeline. If the consultants understand the project business goals, achieve an expedited design review, and comply with the review comments without delay, the project has an excellent chance of being released for sale as scheduled.

You should use the initial meeting with the design consultant team to outline the design concepts for the project. Ultimately, the design approach will guide the marketing program. Successes in the planning board arena directly relate to knowing the cost of the project and of all technical concessions. Answering questions in a pleasant but direct manner in a public forum and communicating knowledge of the project are key elements to achieving an affirmative vote from the planning board.

Any concession in land development design criteria can effectively lower development costs and provide an edge in the marketplace. To obtain design concessions, approach the issue armed with the multitude of manuals prepared by trade associations and professional groups. Have these manuals available during the discussions with the engineering and planning departments. In most cases, this information will not be well received because the design regulations have been approved by ordinance, and government officials are not likely to change the standards without an overwhelming reason to do so.

During the approval process, be forewarned: Often only adversaries will review and publicly comment on the project. Very rarely does anyone stand up in a public hearing and say, "I like this project," "We should approve this project," or "Let's go forward in providing new houses for the new families moving into our community." Thus, it is both easier and more prudent to take the path of least resistance. Achieve approvals as quickly as possible to limit soft costs and interest costs for the land.

Turning new concepts into a project plan that may or may not be embraced by the home-buying public is risky. As an example, zero lot line housing has a market and is an excellent planning concept to increase density. If the marketplace is not ready for this housing concept, however, the project will suffer. *Be a leader in design planning, but do not be a pioneer for new ideas.* From a cost or marketing perspective, these ideas may prove worth the expense and effort to gain community acceptance. Still, it is rare to find a planning board, board of adjustment, or council that will embrace new concepts for a community without significant public input. To be on the safe side, use conventional design approaches, limit the number of design waivers, and, if at all possible, refrain from any variances that would cause debate at a public hearing. The ultimate goal in the land development business is to find a well-located parcel that has zoning in place, a below-market raw land price, and a pro forma showing moderate market absorption while maximizing the profit margin. If any of these land development rules are compromised, be prepared for a prolonged regulatory debate, delay, and increased soft costs.

During the last 5 to 10 years, the most significant change in the construction approval process has been related to environmental regulations, that is, protection of wetlands, rare vegetation, and endangered species.

Appendix 3-6 is a comprehensive regulatory checklist to be used in getting to know the jurisdiction governing the project from the land purchase through the closing of the last lot. Create and maintain a contact file of names, titles, positions, phone numbers, fax numbers, and even the secretary's name for each regulatory department. The administrative and secretarial staff are involved in the day-to-day processing of the project plans. Make their jobs easier by making sure the submission applications are 100% complete. Depending on the project size, absorption, and other factors, most housing projects take several years to complete. Therefore, business relationships with government officials are key to project success. As the purchaser of the property, you should go personally to each regulatory department and discuss the specifics of the project with the staff. Do not delegate that responsibility. The more you know about the land, the better your chances of successfully developing raw land into finished lots.

Value Analysis

Real estate success depends on *location, location, location.* In land development, the cost of raw land and the cost of approving and improving land for the ultimate user will determine which location is best suited to meet business objectives.

Use Appendix 3-7 to summarize basic characteristics for each site you are considering and as a general guide to determine which site has the greatest potential for housing. Identifying reasons for buying or not buying the land requires you to be objective. List pros and cons with an unbiased perspective. As reasons for buying the land, list all the site specifics: location, topography suited for minimizing earthwork, easy road access, excellent vegetation, and so forth. Listing the reasons for not buying the land ensures that the negative viewpoint, if any, has been adequately stated. Every piece of property has a skeleton in its closet, and the job of the site evaluator, or person so charged, is to find the hidden problems. The possible pitfalls must be highlighted before you can fully evaluate the development potential of the site. If you know the history of the land, then you can decide how to overcome problems and you can identify the solutions. Many failed housing projects can be attributed to the developer paying too much for the raw land at the very start.

Land price comparisons in terms of per acre or per lot are an excellent way to compare yields for alternative sites. Most property owners rate the value of their property much higher than the fair market price as calculated by a builder/developer; this stems from the owner's sentimental attachment to the property. The site analysis and market study are useful tools when you discuss property values with landowners. The regulatory process and housing development standards should be explained in detail to the property owners to assure them that the land price is consistent with approved and unimproved subdivisions. As a result of increasing government regulations, the lack of financial commitment to the land development industry, and the competitive nature of regional builders, many projects proceed through partnerships with landowners.

Check comparable land values associated with each site under consideration. Do the research. Orient your project toward the right housing market. Anticipate the time schedule for site selection, governmental approvals, clearing of the first tree, and closing on the first house. Furthermore, understand whether the project will be released in an overdeveloped market or a market striving for finished lots. This estimation of time is just as critical to business success as is buying the land at the right price.

Investing capital during the site feasibility aspect of the development process is necessary to retain professionals who can assist with the technical aspects of the property. It also gives a certain level of comfort to the developer when a nonbiased opinion of the property's potential can be presented with the pluses and minuses of each site. Once the facts are outlined and conclusions reached, proceed toward controlling the property for company use.

Contract Provisions

When you are attempting to purchase property, your goal should be to hold onto the land as long as possible before closing. Retain the option on the land while obtaining all governmental approvals and permits, and relinquish as little cash as possible to hold the land. Make sure that the first phase of the project can obtain governmental approval quickly and that the cash flow requirements can be properly managed. Use Appendix 3-8 to highlight items requiring negotiation between the property owner and the buyer.

All the information obtained during the market study, the regulatory analysis, the site review, and the value analysis will be used in negotiating the final contract for the land. Similar to a land development contract, the raw land purchase contract should provide an escape clause, provisions, and conditions that provide an avenue to void the deal.

Appendix 3-9 presents a land purchase contract outline summarizing the specific items to be included in the purchase agreement. The contingency clause, which allows the contract to be terminated with cause and without financial loss, must be included in the purchase agreement. Raw land purchase contracts should be prepared by an attorney hired by the purchaser. Most raw land purchases are complex, and legal representation is important to execute the contract provisions consistent with all applicable laws.

Appendix 3-1

COMPETITIVE HOUSING MARKET PROFILE

The housing research development status report, market profile, and design features report will clearly define the market and assist in making the decisions on what to build and sell. Company goals will dictate the market segment needed to complement the housing project inventory of the company. The selection of a starter market with fast absorption or a custom home market with a slow unpredictable absorption may result in the same profitablity over time.

HOUSING RESEARCH DEVELOPMENT STATUS REPORT

This report will identify which projects will have product and will indicate the level of inventory within the marketplace. Use this information to judge the market segmentation of the product line.

A. **Development status report** will describe:
   1. Number of planned units in each project
   2. Units completed - sold or unsold
   3. Units under construction - sold or unsold
   4. Units completed - sold or unsold
   5. Units remaining available for sale

B. **Product design features report** will describe:
   1. Master Bedroom
      a. Size
      b. Location
      c. Separate dressing room
      d. Separate sitting room
      e. Ceiling features, such as volume, tray, etc.
      f. Walk-in closets
      g. Fireplace
      h. Flooring
      i. Other
   2. Master Bedroom Bath
      a. Separate tub and shower
      b. Whirlpool tub
      c. Double vanity
      d. Vanity top material
      e. Manufacturer of faucets
      f. Enclosed water closet area
      g. Flooring
      h. Other

3. Kitchen
    a. Style
    b. Counter island
    c. Separate pantry
    d. Separate breakfast nook
    e. Eat-in kitchen
    f. Access to garage
    g. Access to laundry room
    h. Appliances offered as standard
        1) Refrigerator
        2) Dishwasher
        3) Garbage disposal
        4) Oven
        5) Range
        6) Microwave oven
        7) Number of sinks
        8) Other
    i. Cabinets
        1) Style
        2) Material
    j. Counter tops
        1) Style
        2) Material
    k. Flooring
    l. Other
4. Exterior
    a. Finish (i.e., vinyl, wood, stucco, etc.)
    b. Veneer (i.e., stone, brick, etc.)
    c. Roof shingle
    d. Windows
    e. Garage doors
    f. Shutters
    g. Other
5. Interior
    a. Number of bedrooms
    b. Number of baths
    c. Covered porch
    d. Center hall foyer, clear story
    e. Formal foyer
    f. Air conditioning
    g. Ceiling heights
    h. Fireplace
    i. Flooring
    j. Security system
    k. Number of car garages
    l. Basement
    m. Patio or deck
    n. Other

Appendix 3-2

SITE EVALUATION FACTORS

Factor to Be Evaluated	General Source of Information
Boundary survey and acreage	Deed, tax maps, previous surveys, tax offices, restrictions running with the land such as easements
Zoning: lot, building restrictions	Zoning maps, local zoning board, building department
Site improvements and planning policies	Local planning department, subdivision regulations, county planning office
Topography	United States Geological Survey 7.5ft quadrangle maps, local planning department, field surveys, aerial maps, local land owners, library
Soils	Soil survey prepared by the United States Department of Agriculture Soil Conservation Service, on-site subsurface exploratory borings or test pits, contact septic installers and contractors
Floodplain and drainage	Reports prepared by the United States Army Corps of Engineers, state or local department of environmental protection, FEMA, engineering and road departments: public works departments
Water system	Local water authority or water company, fire marshall
Water from wells	State geological survey, state health department, local approving authority, well drillers
Sanitary sewer system	Local sewer authority, state water control board, engineering departments

Factor to Be Evaluated	General Source of Information
On-site sanitary system	State health department, local approving authority, septic installers
Freshwater wetlands	State wetlands mapping, local planning offices, field survey
Adjacent or nearby subdivision plans	Local planning and engineering offices
Roads, traffic, and access Future road improvement plans	Local planning departments, state department of transportation, engineering departments
Hydrology/geology, aquifer recharge	State geological survey, local planning department, contact well drillers
Vegetative cover	Site inspection, aerial photographs
Gas and electric	Local power company, state public utility commission
Telephone	Local telephone company, state public utility commission
Archeological/historical significance	State historical society, library
Storm-water management	Local county engineer or department of public works, local planning office, state or regional agencies
Sediment and erosion control	Local department of public works, soil conservation service
Parks/recreation	Local or state parks planning office or recreation board
Schools	Board of education
Landscaping	Local planning department, local nurseries, library, landscape contractors
Transportation availability	Bus, rail, airports for commuting
Public services	Police, fire emergency, trash pick-up
Shopping	Malls, strip centers, local chamber of commerce

Appendix 3-3

LAND EVALUATION SCOREBOOK

Parcel Lot and Block______________________________ **Date**________

Location____

Factor	**Assigned Point Totals**				**Score**
	1	2	3	4	
Distance to Major Population Centers	30mi	20 mi	10 mi	5 mi	_______
Distance to Interstate Highway	10 mi	5 mi	3 mi	2 mi	_______
Distance to Nearest Shopping	5 mi	4 mi	3 mi	2 mi	_______
Distance to Nearest School	5 mi	4 mi	3 mi	2 mi	_______
Distance to Nearest Public Services	5 mi	4 mi	3 mi	2 mi	_______
Distance to Nearest Competition	5 mi	4 mi	3 mi	2 mi	_______
Condition of Road Frontage	Poor	Fair	Good	Excellent	_______
Topography	Low	Flat	Rolling	Hilly	_______
Extent of Slopes Over 10 %	75–85	50–75	25–50	Under 25%	_______
Woods	Poor	Fair	Good	Excellent	_______
Soil	Poor	Fair	Good	Excellent	_______
Rock	Extensive	Frequent	Occasional	None	_______
Natural Gas	3 mi	2 mi	1 mi	At site	_______

Factor	Assigned Point Totals 1	2	3	4	Score
Electric Power	3 mi	2 mi	1 mi	At site	______
Public Water Supply	None	500–1,000 ft	100–500 ft	Within 100 ft	______
Public Sewers	None	500–1,000 ft	100–500 ft	Within 100 ft	______
Wetlands Portion of Site	Extensive	20%	10%	Under 5%	______
Approximate % of Land Available for Lots	Under 25%	50%	75%	100%	______
Single-Family Zoning	1 acre	1 acre	½ acre	¼ acre	______
Subdivision Standards	Excessive	Average	Limited	None	______
General Impression of Site for Single-Family Homes	Poor	Not First Choice	Further Investigate	Great	______

General Comments:____

____

____

____

Rating of Site________________________________

Prepared By/Date ________________________________

A site with a rating of over 55 should be further evaluated for development.

1 ft = 30 cm; 1 mi = 1.6 km; 1 acre = 0.4 ha

Appendix 3-4

SITE ANALYSIS CHECKLIST

REPORT BY:______________________DATE:_______________
LOT:_________BLOCK:___________LOCATION:____________

1. Highlight special land characteristics that would cause a "NO" purchase decision and indicate any cost implications.
   a. Poor soil
   b. Topography
   c. Heavy vehicular traffic
   d. Heavy air traffic
   e. Vegetation
   f. Noise problems
   g. Odor problems
   h. Wetlands
   i. Condition of adjacent property
   j. Poor surface drainage
   k Flood hazard
   l. High water table
   m.Unsightly views
   n. Demolition requirements
   o. Environmental problems
   p. Illegal dumping on-site
   q. Railroad tracks nearby
   r. Interstate highways nearby

   Other:____
2. List soil classification
   a. Advantages
   b. Disadvantages
   c. Depth to seasonal high water table
   d. Summary of soils report for the site

   ____

   ____

   e. Will soil conditions cause excessive development or building costs?

   ____
3. Topography
   a. Source of the topography survey:__________
   b. Date of aerial topography flight taken:__________
   c. Topography map checked by a field survey? Date:_________
   d. Will heavy vegetation or steep slopes effect the topo accuracy?
   e. Examine USGS map and list any discrepancies in the topo________

   ____
4. Buildable area
   a. What percentage of the property is
      low___flat___rolling___hilly___
   b. Indicate on the survey any areas to be excluded as buildable area.
   c. Are there any cost implications as a result of the topography?________
5. Boundary Survey
   a. Are the exact boundary lines known and verified?
   b. Are there any property line discrepancies such as overlaps, etc.?______
   c. Is the seller responsible for resolving title problems?_________
   d. Boundary survey is available by date:______
   e. Summary of discussions with the surveyor of the property

   ____

   f. If the property has not been surveyed, will the seller pay for the work?
   g.Surveys available for the adjacent properties____________________
      What are the impacts of the adjacent property on the site?_________

____

What are the cost implications?________________________

6. Does the land have enough natural fall for drainage purposes?_______________
   a. What is the effect of the topo on earthwork operations?_______________
   b. Is there drainage off-site that drains onto this site?__________________
   c. What are the cost implications?__________________
7. What % of the property might be considered nonbuildable due to:
   topography__________woods___________soil_____________
   rock______________wetlands__________easements________

   What is the net buildable area to be used for residential development?__________

Based upon the density factor of the zone, calculate the number of lots based upon the net buildable area:_____

8. Vegetation growth is: Light___Moderate___Heavy____
   Are there tree-save programs required for this site?________________
   Cost implications for clearing operations or tree-save programs.________________
9. Any natural or man-made features on site: streams___lakes___ponds____
   rivers_____storm-water management facilities___flood zone____
   Should any of these features be used as a site amenity?_____________
   What are the cost implications for making the above features into site amenities?______________________________________
10. Does the property have road frontage? in ___________ft.
   Public or private road access?____________________
   Existing right-of-way width of frontage road___________
   Dedication of right-of-way required along the property frontage in____ft.
   What is the road surface of the frontage road?__________________
   Frontage road is owned by: Federal____State____County____City____
   Private_____Other___
   If private, what are the ingress/egress rights assigned to the site?
   ____
   List future improvement plans of the off-site roads near the site:
   ____
   What are the cost implications to acquire right-of-way access for ingress and egress or for road improvements?________________________
11. Natural Gas Facilities
    - Distance to line
    - Capacity required to serve subdivision
    - New line required
    - Cost implications
12. Electric Lines
    - Distance to nearest lines
    - Relocation of poles required
    - Cost implications
13. Water Lines
    - Distance to nearest line
    - Sufficient capacity to serve the subdivision
    - New line required
    - Need to loop the water lines for the system

Sufficient pressure to serve the property (check with the Fire Department)
Cost implications

14. Private Wells
    Depth_________Quantity Expected_________
    Quality of the water_________________________________
    Treatment required_____restrictions on use - irrigation_____cost____
15. Sewer Lines
    Distance to nearest line
    Sufficient line capacity to serve the property
    Sufficient treatment plant capacity to serve the project
    Alternative sewer systems available, such as septic systems
    Cost implications
16. Other Utilities
    Telephone Company Prewiring cost_______________________
    Cable TV Company Prewiring cost_______________________
    Street Lights required Cost:____________________________
17. Storm Drainge System
    Distance to the nearest storm drain
    Sufficient size to handle flow from the site
    Cost implications
18. List the nature of the adjacent properties:
    North:____
    East:____
    West:____
    South:____
19. Contact List
    Planner:_________________________Phone#__________
    Comments____
    Engineer:________________________Phone#___________
    Comments____
    Soils Engineer:____________________Phone#__________
    Comments____
    Architect:________________________Phone#__________
    Comments____
    Environmentalist:__________________Phone#__________
    Comments____
    Traffic Engineer:___________________Phone#___________
    Comments____
    Surveyor:________________________Phone#__________
    Comments____

20. Site analysis conclusions and recommendations:

____
____
____
____
____
____

Appendix 3-5

TYPICAL RESIDENTIAL APPROVAL PROCESS

A. LAND--**B.CONCEPT LAND PLAN**

A. LAND	B. CONCEPT LAND PLAN
Marketability	*Engineering*
Regulatory Analysis	*Soils*
Site Analysis	*Environmental*
Value Analysis	*Traffic*
	Architecture
	Landscape Architecture
	Land Planning

C. FINAL and PRELIMINARY IMPROVEMENT PLANS

- *Army Corps of Engineers*
  - *Dredge and Fill*
  - *Wetlands*
- *State Environmental Department*
  - *Wetlands*
  - *Flood Plains*
  - *Impact Statements*
  - *Well Permits*
- *Sewer Authority*
  - *Local*
  - *Regional*
  - *State*
- *Water Authority*
  - *Local*
  - *Regional*
  - *State*
- *Municipal*
  - *Planning Board*
  - *Shade-Tree Commission*
  - *Environmental Commission*
  - *Fire Department*
  - *Health Department*
  - *Tax Office*
- *County Planning Board*
- *Soil Conservation District*
- *State Department of Transportation*

1. Post Performance Bonds and Inspection Fees

2. Start Land Development and Obtain Signatures on Final Plat

3. Record Final Plat and Apply for Building Permits

4. Construct Building Pads; Building Permits Issued

5. Land Development 70% Complete; Reduce Bond and Close First House

6. Close Last House; Turnover Infrastructure; Obtain Maintenance Bond

Appendix 3-6

COMPREHENSIVE REGULATORY CHECKLIST

Report By:______________________________Date:______________

Lot:______________Block:________________Location:______________

1.Municipal planning board address:______________________________
Planning board secretary:____________________Phone#:______________
Fax#:__________________

2.Municipal engineer address:______________________________
Engineer contact:________________________Phone#:______________
Fax#:__________________

3.Police Department contact:____________________Phone#:______________
Fire Department contact:______________________Phone#:______________
Emergency Services Contact:____________________Phone #:______________

4. What is the local attitude toward residential development?________________
Planning board members:______________________________
Neighbors:______________________________

5. Municipal architectural review board contact:______________Phone#:__________

6. Obtain all relevant ordinances and indicate cost problems with regulations:
Subdivision:______________________________
Building:______________________________
Zoning:______________________________
Development:______________________________
Water:______________________________
Sewer:______________________________
Master plan:______________________________
Growth management plan:______________________________

7. Present zoning of the site:____________________
Density permitted:______________________
Minimum lot dimensions: Size____________Width__________Depth________
Lot frontage:__________Minimum sq ft of building:________
Setbacks: Front______________Sides______________Rear____________

8. Does the local municipality require subdivision bonds?____________________
Accept? Letter of Credits______Performance Bond________lender's Agreement_____
Cash Requirements__________Bond Amount Calculation__________________
Will partial bond releases be approved?______________________________
Maintenance bonds required?______________Number of years?____________
Manitenance bond amount calculation______________________________

9. Property in Special Regulatory Zone: Forest____Wildlife____Fire Hazard____Fault___.
Historical___Flood___Dump____Acquifer___
Archeological____Environmentally Sensitive____

10. List governmental policies on impact fees, such as schools, parks, recreation, sewers, etc.

____

____

11. List unusual subdivision requirements and potential cost implications:

____

____

12. List basic steps in subdivision approval process including time table:

____

____

____

13. Indicate political climate toward housing projects:

____

____

14. Summarize discussions with

    Planning department:____________________________

    ____

    ____

    Engineering department:____________________________

    ____

    Building department:____________________________

    ____

    ____

    Zoning department:____________________________

    ____

    ____

    Planning board member:____________________________

    ____

    ____

15. Summarize regulatory assessment of proposed site and subdivision goals (pros and cons)

____

____

____

____

Appendix 3-7

VALUE ANALYSIS CHECKLIST

Report By:____________________________________Date:_______________________

Lot:________________Block:________________Location:________________________

1. Comparison of price per acre or price per lot with other sites under consideration:
   Higher (Why):____
   ____
   Lower (Why):____
   ____

2. What is the potential use for this land?______________________________________

3. When would this land be ready to close lots/homes?_____0–1yr____1–2yrs____3–4 yrs

4. Is the price of the land low for the land to appreciate from inflation or speculation?____

5. What did similar property sell for: 1 yr ago________3 yrs ago_______5 yrs ago______

6. What are the land price trends for the area in the future?__________________________

7. How long can this land be held in inventory while approvals are obtained?___________

8. What is the raw land per lot objective for this site?_____________________________

9. Will the price objectives be met once the project obtains all regulatory permits and approvals?____

10. Why buy this site? Investment_____Design, Approve and Sell_____Build Out______

11. Will this property be used for our own purposes?______________________________

12. Outstanding reasons for buying this land:____________________________________
    ____
    ____
    ____

13. Outstanding reasons for not buying this land:________________________________
    ____
    ____

14. **PROCEED**________________________**REJECT**__________________________

Appendix 3-8

LAND PURCHASE CHECKLIST

Report By:______________________________Date:____________________
Lot:______________Block:______________Location:____________________

1. Size of parcel______________ac. Zoning____________Lot size______________
   Number of lots based on zoning requirements and gross acres________________

2. What portion of the site has been excluded in computing net buildable area?________
   Net buildable area:________________
   Number of lots based on net buildable area:______________________________

3. Any unusual cost considerations: On-site_____________Off-site________________

4. Present offer based on: Acreage____
   Net buildable area__________________________________
   Number of net lot yield______________________________
   Comparable land values______________________________

5. **Asking Price**______________________**Offered Price**______________________
   Cash down payment and terms asked?__________________________________
   How much land to be released for down payment?__________________________
   How firm is the seller on price and terms?______________________________
   Is the property reasonably priced?____________________________________

6. What factor would cause the price and terms to be negotiated?__________________
   ____

7. What terms should be offered?____

8. Can portions of the land be released without full payment?_____________________

9. Indicate contingency clauses to be included in the purchase agreement. These contingencies are concerns that would cause the contract to be terminated with cause.

Soils Report	_____	FHA/VA Approval	_____
Engineering Study	_____	Environmental Study	_____
Final Plat Approval	_____	Water /Sewer Allocation	_____
Number of Lots	_____	Availability of Utilities	_____
Access	_____	Easements	_____
Zoning	_____	Annexation	_____
Condemnation	_____	Recordation of Plat	_____
Building Permits	_____	Other	_____

Appendix 3-9

LAND ACQUISITION CONTRACT OUTLINE

1. General: Buyer and Seller
2. Mutual Agreement to Buy and Sell
3. Description of Property
    a. legal description
    b. address, lot, block, municipality
    c. subdivision plat
4. Purchase Price
    a. amount: lump sum, per lot, per acre
    b. payment: method and terms
    c. security for unpaid balance
    d. releases
    e. subordination clauses
    f. optional clauses
5. Warranties of Seller
6. Contract Contingency Clauses
    a. zoning
    b. building permits
    c. regulatory permits
    d. water or sewer availability
    e. other clauses
7. Inspections of Property
    a. engineering studies
    b. soils investigation
    c. environmental assessment
    d. topographical and boundary surveys
8. Title Objections
9. Closing: Time
10. Condition and Possession of the Property
11. Default Provisions
12. Earnest Money
13. Commissions
14. Conditions of Closing
15. Assignments
16. Notice Provisions
17. Survival of Contract
18. Special Stipulations
19. Execution of Contract

Chapter 4

Due Diligence

General

Due diligence analysis of vacant property is used by land buyers to assess the property for their intended use: land investment, entitlements, land development, or vertical construction. The essence of a due diligence report is to document what is known about the property. This report will assist buyers and sellers of real estate for development purposes. With all the "good land" developed, the land remaining has something wrong with it. *Good land* is defined as land void of regulatory constraints. It may have physical constraints and land form characteristics that make development difficult, but land development engineers find ways to resolve these issues.

The federal and state governments own about 39% of the land in the United States, about 60% of the land is privately owned, and 1% is foreign owned. The US land area totals nearly 2.3 billion acres with less than 5% developed for residential use. It is essential then to develop a step-by-step site analysis and due diligence program and follow it without exception. Assessing land is all about finding everything that is wrong with the land rather than what is good about the property.

Urban sprawl is a misleading representation of development patterns. There, development will leapfrog over land areas for many reasons:

1. Seller does not want to sell.
2. Seller wants to sell, but the price is too high.
3. Public water and sewer are unavailable.
4. Site constraints limit the buildable area.
5. Government zoning and/or political problems must be resolved.
6. Infrastructure to support new development is lacking.
7. Parcels of land are too small or there are strange property line configurations.
8. Endangered species are present.

And on and on!

Due diligence reports are undertaken by developers to minimize the risk and to identify cost implications of each issue discovered during the due diligence period. Resolving any issue will have a direct bearing on the decision to purchase or walk away. Time and money expended to resolve issues may not be enough to overcome these issues. Buyers must not let emotion rule the decision, and sometimes it is better to leave money and time on the table than move forward with a purchase.

As with any report, the initial section will be to outline the general overall description of the property. The following information would be essential components of the "General" section of the due diligence report.

Property Description and Ownership

Include name, address, city, phone number, and email address: A complete contact list for the owners of the property is a great reference. The property description will be a location description or directions to the property.

In many cases, property not developed and located close to an urban core is vacant primarily because the seller is asking above-market prices. Site analysis conclusions can be used to negotiate better prices.

Definitely find out if there are separate deed restrictions recorded that encumber the property.

Legal Description

The legal description is important for researching the property in public records. For this purpose, a lot and block, or parcel identification number would be sufficient. Try to obtain a metes and bounds description of the property from the owner or a boundary survey.

The boundary survey would show easements, rights-of-way (ROW), and/or encroachments, all of which affect the upland area suitable for development purposes.

Tax Map and Parcel Size

The local municipal tax records/maps will have references to include in the report. The parcel size is typically stated in acres.

Master Plan Compliance

Visit your local municipal government planning office and find the parcel on the master plan approved for the community. This will outline government's future overall growth patterns for the subject property and surrounding areas. If the intended use of the property is consistent with the master plan and zoning map, the project may be considered a use-by-right project and the approval process will be much easier to navigate. However, most master plans are updated in 10-year intervals and amending the master plan may be the only way to develop property consistent with surrounding uses.

Future Road Alignments and Rights-of-Way Dedications

Obtain from local government their plans for new roads, road realignments, and expansion of road rights-of-way. Determine how their plans affect the subject property and the surrounding area.

Review of Phase 1 Environmental Study

If a Phase 1 environmental study has been prepared for the subject property and the owner provided a copy, review the findings in detail. This report, if not available, would be required by lenders and is essential from the buyer's perspective. If the Phase 1 report identifies issues, a Phase 2 report is required. This analysis will dive deeper into the issues raised, and the expense may be significant to resolve an issue. This, of course, is also a negotiating tool to be used to the buyer's advantage.

Site Pictures

Walk the site and take enough pictures to document the site characteristics, adjacent properties, and existing roads. Include the essential overall site photos in the report. An aerial photograph is very important in assessing how the property is situated with the overall growth pattern of the immediate area. Check public and online sources first; if not current, retain an aerial company to fly the site and take sufficient photos from several perspectives. Often, Google Earth provides an excellent option at a fraction of the cost of an aerial photograph.

Maps and Plans

With today's technology and the need for local government to keep pace with the information flow and development changes in their community, most local municipalities post their ordinances and maps online. Compile maps and plans available online and locate the subject property on each map or plan. Check online resources for GIS mapping technology with topography and aerials for updated information. Some plans can only be obtained from a regulatory agency, that is, water and sewer line as-builts, but it is worth the effort. Analyze the property location and the site characteristics to determine the development potential.

The following list of map or plans are invaluable in the site analysis work required in land acquisition.

Property Boundary Survey

At this point in the site analysis work, you may not have contacted the property owner. Thus, visit the municipal engineering or public works office and ask for surveys for the subject property or adjacent properties. There will be surveys available in and around the subject site unless the site is located outside the infrastructure core. This survey will illustrate boundary line problems, that is, cores, overlaps, access easements, rights-of-way, utility easements, and so forth. The results will show the constraints to developing the site. While at the government office, request a copy of the road construction plans for the property frontage road.

Tax Maps

This is an important overall map showing how the subject parcel fits like a puzzle with surrounding property. The tax map provides the lot number, blocks, parcel ID numbers, and other references that can be used for deed research, tax information, and prior ownership history. This map needs to be studied for development and ownership trends. Disney World land was massed by Disney using the tax map to find separate landowners and quietly purchase property to assemble a large contiguous land area for the theme park.

Advocate rezoning for value. Simply stated, buy land with a less valuable use classification and rezone the property to its highest and best use. As an example, buy agricultural use property for land located within a commercial area. If the rezoning to commercial is successful, the value of the property has increased in value. The real estate development business is a risk reward industry. To assess the possibility of rezoning for value, the zoning map and master plan are important public documents to be researched.

Zoning Maps

This map is posted online by the local municipality. The map will show all the zoning districts within the municipal boundaries. Locate the subject parcel on the zoning map and determine the assigned zoning classification for the property. Also, make note of the zoning classification for the immediate surrounding and adjacent properties. Assess trends and changes in zoning for the surrounding properties. Also, contact the zoning or planning department and seek information about their intention of revising the zoning map for the town.

Master Plans

Each jurisdiction will have an overall growth management plan illustrating the uses for areas and property within its jurisdiction. This plan will delineate uses, that is, residential, industrial, commercial, and so forth. The master plan does not specify the zoning but designates the use. The master plan changes periodically; in some cases, it must be updated every 5 or 10 years based on local ordinances or state law. The property should be consistent with the master plan or consider a master plan change application. This may present a more difficult and lengthy regulatory process, but most jurisdictions permit this option.

Street Maps and Community Amenities

Find the best possible map to show the benefits of the community: features, schools, shopping, offices, employment centers, highway system, parks, and so forth. Locate the property on this map for future reference and distribution. The best resource is the chamber of commerce.

Topographic Survey or GIS Maps

The more advanced municipalities will have flown topography maps for their jurisdiction. Topo from an aerial is much more reliable than GIS. In either case, the engineering office at local government will have topography maps for your use. Buy the necessary topo maps of the property in question (PIQ). Ask for adjacent property topography and adjacent development plans, if available. Also, ask for the construction plans, if any, for the frontage road servicing the property. Evaluate the information for consistency in elevations, land features, and property lines. Review low areas for wetlands, floodplains, jurisdictional streams, and/or poor soil. Delineate areas not suitable for development purposes. Severe grades, excessive nonbuildable areas, or road frontage issues may hinder a cost-effective development approach.

Do not force the issue but be exhaustive in the analysis of the information. You are determining the priority of the site for development potential.

Without much of a debate from an engineering or construction perspective, a field survey is preferred over GIS and aerial flown topography. This is especially true if the local jurisdiction requires a tree survey. The tree survey requirements could include identifying and mapping all trees more than 12 inches (in.) in diameter at breast height (DBH).

Aerial Photos

A recent aerial photograph provides information not readily understood via plans, maps, or site visits. The aerial provides property use relationships and the overall development approach of the immediate surrounding area. Vacant parcels are also easily identified (Figure 4-1). If the location selected is Priority A, the cost of an aerial photo will vary, but a recent aerial in Charlotte, North Carolina, with labels cost less than \$500. Most online aerial photos lag development, but with the downturn, aerial photos may catch up with development activity. If the site is infill, online aerial photos should be sufficient. This element of the due diligence package is essential to the overall analysis of a site.

Soils Maps

The soil conservation district maps delineate the soil characteristics for every county. These maps provide an overview of the types of soil found in a location. The maps are dated by use of old aerial photos with soil types delineated. The key is to locate the PIQ on the aerial, sometimes not an easy task. Use a detailed road map to assist in this effort. After locating the property on the soils map, record the soil type on the property and use the soil conservation district soil descriptions and classifications to assess the positive and negative aspects for development operations.

Figure 4-1. Aerial photos are the best way to assess the relationship of the PIQ with surrounding and adjacent properties. Other mapping and aerial software and search engines make it easy to assess views, site constraints, and frontage issues.
Source: David E. Johnson.

As an example, the report may indicate a high-water table for a specific soil type. This may impact development costs, because it may alter the building foundation design, lead to classifying the soil type as wetland, or indicate that the cite has rock 10 in. below the surface. Study the map and the soil types. This document is an overview and should be followed up with a strategic soil investigation program on site by using test pits and soil borings. The on-site investigation will further define the soil types and assist in delineating the extent of the soil types found on site.

Water and Sewer Maps

These maps can be obtained from the municipal water and sewer department, or your community could be served by private companies. In either case, plans illustrating the location of existing water and sewer lines exist in some fashion, if only for maintenance purposes. Also, seek to obtain capital improvement plans showing future water and sewer extension projects. It is easier to follow water and sewer, but the land cost may reflect the availability of the utilities. Ask for a paper printout for your future reference. For water distribution, the size of the main and its location plus fire hydrant locations is sufficient for the initial assessment. For the sewer line, location of the line, manhole locations, line invert rim elevations, and sewer line size are essential to determine if the PIQ can be served by gravity sewer. Note: the elevations shown on the municipal plans must be in the same datum as your topography map to make a quick assessment of whether the property may be served by gravity sewer.

Wetland Inventory Maps

Many states have used soil conservation maps or GIS maps to overlay wetland areas and jurisdictional streams for delineating environmentally sensitive areas. This illustration is not precise. However, in evaluating the development potential of property, these maps provide guidance on the location and size of area that must not be disturbed and thus is not considered to be buildable area. Historically, the majority of remaining property has been developed. The remaining land most likely includes wetlands or jurisdictional streams. Your state EPA is the agency to contact for copies of the wetland maps.

Compare the wetland areas or jurisdictional stream locations with the soils map. The soils in the area of wetlands or streams should be classified as unsuitable soils for construction activities.

Floodplain Maps

FEMA has posted the flood insurance rate maps on its website. By entering a property address, the flood map panel will be located for review. This map illustrates the approximate floodway and may or may not include the floodplain elevation at cross sections depicted along the stream. The *floodway* is the stream channel and any adjacent areas that must be kept free from encroachments to allow the 1% annual chance of flood carried without an increase in elevation. A floodplain shown on a property will affect its development potential, increase permitting requirements, require significant design considerations, and reduce the buildable area suitable for residential and nonresidential uses.

Road Network Plans

Many communities prepare a comprehensive road network plan illustrating transportation links within a jurisdiction to accommodate future development. These road plans are anticipated road alignments usually locating road extensions, realignments, or new roadway systems. The roadways provide a transportation plan used by government in reviewing future development plans. The local or county government will have a road alignment plan depicting the future road network serving future development. It is possible a future alignment will impact the development potential of a site by either accommodating the right-of-way in the design plans, improving off-site roads, or contributing to the traffic study for further analysis based on the development impact of the project. This plan may have the most significant impact on a project design, especially if it is overlooked during the due diligence period.

Zoning and Land Uses

Zoning regulations and restrictions are used by municipalities to control and direct the development of property within their borders. There are more than 80,000 local municipalities in the United States, each with its own unique approach to controlling how land is used. Most will have an updated zoning map outlining the specific zoning classification for each parcel of land within its boundary. Remember, review the zoning map for inaccurate representations of actual development patterns and focus on those properties for which a zoning change would add value and be justified for local politicians to accept. In the previous section, a land-use plan may designate a specific area for residential use. Within the broad land use of residential, the zoning would be, for example, single-family detached homes on lots with a minimum of 10,000 ft^2. In this example, the specific bulk requirements governing the development pattern of each lot would be defined in the zoning ordinance. Each municipality has its own zoning classification or label for each specific one, and the bulk requirements may differ.

To illustrate rezoning for value, for example, answer these questions:

1. How much does residential zoned acreage sell for in your local area (before the recession and currently)?
2. How much does commercial zoned property sell for per acre?

3. How much does it cost to rezone property from residential to commercial?
4. What is the time frame to accomplish the rezoning?
5. Does it make sense from planning, marketing, and financial perspectives to change the zone?

Go to your local planning office or go online and study the zoning map for background knowledge and understanding of the local development patterns. In the future, zoning may become a much more important tool for local government in controlling development patterns within their boundaries.

Permitted and Conditional Uses

For each zoning classification, there are permitted and conditional uses defined by local government. As an example, for a property zoned general business, the following permitted and conditional uses guide the development pattern approved by local government.

Permitted Uses

From the zoning regulations, determine the permitted uses for the property. Determine if any of the uses are consistent with your intended use for the property. Evaluate the adjacent property for consistency in development pattern and if this zoning is appropriate for the area or determine whether there is a higher and best use for the site.

As an example, for a general business use, the permitted uses would be outlined as follows:

1. All principal uses permitted in the neighborhood business zone,
2. Art, dance, music, or other similar instructional school,
3. Auto parts store,
4. Bowling alley,
5. Commercial office,
6. Commercial retail,
7. Funeral home,
8. Garden center,
9. Professional office,
10. Shopping center,
11. Vocation/trade school,
12. Combination of two or more of the preceding permitted uses, or
13. Other uses similar to those already listed.

Conditional Uses

Zoning ordinances include other uses which can be developed on the property but must not contain any variances and must be approved. An example of conditional uses for a general business zone are the following:

1. Automobile repair shop,
2. Automobile sales establishment for new or used cars,
3. Automobile service station (gas station or motor vehicle service station),
4. Car wash,
5. Churches and other places of worship,
6. Commercial recreation activities,
7. Health care facility,
8. Hotel or motel,
9. Mini-storage facility,
10. Public utilities, or
11. Veterinary clinic hospital or animal care facility.

The preceding examples are from an ordinance that has not been updated in years. In the future, mixed uses will be the only way to develop sustainable projects. Both vertical integrated mixed-use buildings and horizontal mixed uses provide market flexibility and diversification to reduce risks. However, these types of projects will necessitate new partnering roles and the reinvention of business models within the real estate development industry. Remember, a sustainable project achieves economic, social, and environmental objectives. A project, product, or building labeled "green" may not be considered sustainable.

Zoning Restrictions, Bulk Requirements, and Constraints

Each zone has certain restrictions imposed to guide development patterns on each specific property. As an example, a residential land use with R-4 zoning classification allows four lots per acre, or it may be defined by lot size, that is, 10,000 ft^2 minimum.

The bulk requirements listed next are shown on the single-family lot illustration. If an applicant or the property owner wants to change any of the bulk requirements, that is, change the minimum lot size to 8,000 ft^2 minimum, government must approve of this change through a variance procedure. In many overdeveloped areas, variances to the bulk requirements are difficult if not impossible to obtain. In growing communities, government officials will at least listen to and approve or disapprove the variance request on its merits. It is incumbent on the applicant to make the case for why the variance should be approved by local government.

A current trend is for applicants to seek conditional zoning approvals rather than a straight rezoning. The conditional zoning allows development flexibility so builders can react to market changes. The conditions of the approval are developed between the builder/developer, politicians, and municipal staff.

Subdivision Bulk Requirements

Typical constraints on property are easements or rights-of-way imposed on the property resulting from various actions by the landowner or government. As an example, a single-family residential zone may have a 10 ft wide shade tree or utility easement along the front property line. This easement may restrict the property owner from using this 10 ft wide area for his or her enjoyment and use. Many landowners have sold rights-of-way to utility companies for the installation of overhead high-tension power lines. The width of these easements vary but may be as wide as 300 ft and traverse the property at any angle. Not only do these rights-of-way restrict the use of the property, but the development pattern is restricted and the marketing potential of the land is diminished.

Minimum lot size:	10,000 ft^2
Maximum density:	4 units/ac
Minimum lot width:	75 ft
Minimum lot frontage:	25 ft
Minimum lot depth:	100 ft
Front setback:	25 ft
Side yard setback:	10 ft
Minimum rear yard setback:	25 ft
Maximum project impervious surface:	25%

Residential developments are minimizing lot sizes and maximizing the building footprint. Side yard setbacks will become an issue as they relate to building and fire codes.

Other easements that restrict the use of property may be for the installation of public utilities, that is, sewer lines, water lines, storm drainage systems, or ingress–egress. As outlined previously, floodplains, jurisdictional streams, buffers, and wetland areas will also be illustrated on the property survey and restrict the development pattern of the property.

Density Bonus

An applicant may be offered a density bonus. This is an incentive-based provision allowing applicants to increase the maximum lot yield in the project for concessions that would assist the community in achieving policy goals such as affordable housing. It is also a win–win if a density bonus is associated with cluster zoning. Cluster zoning permits smaller lot sizes to consolidate infrastructure and promote larger open space areas within the project boundaries.

Yield Analysis, Open Space, and Buffer Requirements

Open space and buffers take "upland" area residential zoning and establishes a number of units per acre permitted on the property. The acreage is the gross acres of the site without any reductions. To determine an estimated number of lots before spending the money and time to prepare a land plan, use the following approach to determine the yield (adjust percentages according to the local government regulations, that is, 20% for right-of-way is for a typical 50 ft right-of-way width):

Example

Determine the gross acreage of the site, such as 36 ac.

Reduce the gross area for other site constraints that would limit upland area or restrict development, that is, floodplains, wetlands, jurisdictional streams, steep slopes, boundary overlaps, future road alignments, and so forth.

Estimate land area to deduct from the gross acres, such as 7 ac.

Subtotal upland area then equals 29 ac.

Reduce upland area for land used for public rights-of way (use 20% of upland acres), 5.8 ac.

Reduce upland area for land used for SWM/BMP (use 10% of upland acres), 2.9 ac.
Reduce upland area for land set aside as open space (use 10% of upland acres), 2.9 ac
Reduce upland area for any land dedicated for frontage road right-of way, e.g., 1 ac.
Total upland area for development purposes then equals 16.4 ac.

Use the density associated with the property zoning:
(e.g., 3 units (du)/ac and multiply the total upland area, that is, 16.4 ac × 3 units/ac) equals 49.2 units
Use 49 units for financial analysis.

Regulations

To investigate the development potential of a site, research local, regional, state, and federal regulations and determine thc impact these regulations will have on developing the site. Such regulations include

- Rezoning applications;
- Development impacts on services and existing infrastructure;
- Dedications, easements, and encroachments;
- Road access, off-site road improvements, and on-site road circulation;
- Land development regulations;
- Stormwater management facility requirements;
- Best management practices;
- Regional, state, and federal regulations;
- Community "hot buttons" (local unwritten regulations); and
- Entitlement, approval, and the permitting process.

Assess the complexities of the regulations and calculate how the regulatory climate will affect the following:

1. Purchase, financing, and market entry by the time required to obtain all permits and approvals;
2. Costs and project pro forma;
3. Net development area of the site; and
4. Layout of the property for the intended use (use 10% of upland acres).

First and foremost, concentrate on local regulations because every municipality controls real estate development differently. A municipality will stress certain regulations based on their experience level, and many of those requirements may still be open for interpretation. This section of the due diligence report highlights current technical conditions imposed by government that may or may not affect the development potential of a site.

Rezoning Applications

During the site analysis phase, adjacent properties will be researched for proposed uses and assessed for the quality of the current use, design, and condition. Adjacent properties will impact how the PIQ is developed.

Visit the local planning department office and meet with the senior planner. The goal is to review current rezoning applications in town and review any past rezoning cases that match the development program for the PIQ. Also, review any rezoning applications approved or denied on any property with a similar use as the PIQ and certainly for any properties in the immediate area. This is important to understand the local government's view on development patterns and what conditions are imposed on applicants.

Discussing the rezoning cases with the planning staff will give insight on how the regulators think and how the local politicians react to a development proposal. The goal of any project is to obtain local approvals from regulators and politicians. Understanding their perspectives on developing land is essential to the overall success of the proposed project. Including this information in the site analysis report will provide valuable insight for the client to make an informed decision.

Development Impact on Services and Existing Infrastructure

Most residential and nonresidential projects have an impact on existing municipal infrastructure and services. In most jurisdictions, the size of the project in area, units, or building square footage proposed will trigger the need for impact studies. The intent is to assess government service areas and the existing infrastructure and its capability to absorb the development potential of the project. An assessment of services may include fire and police protection, solid waste pickup, schools, recreation elements, and so forth.

The existing infrastructure, that is, roadways, intersections, water availability, sewer lines, and drainage facilities, should be assessed to determine if the proposed project will affect the infrastructure. To provide a remedy to adverse impacts, developers will pay for improvements or contribute their fair share to any improvements required as a result of the project. The intent of this analysis is to define and budget for anticipated costs during the due diligence period.

Dedications, Easements, and Encroachments

Research the land development ordinance for requirements associated with the dedication of additional land for the existing road rights-of-way that border the property. Also, understand the width of easements required for water, sewer, and drainage facilities that extend beyond the public right-of-way. The easement width will be an encumbrance on the property or lot. For a single-family lot layout, the width of easement should require an evaluation of the lot width with the easement. The goal is to place easements along property lines, and the width of the easement on each single-family lot should be less than the side yard setback. If a boundary survey is available for the property, review the property lines for any encroachments from adjacent landowners, such as fences and structures. Review the encroachment(s) with the local zoning official and determine an approach to correct the problem.

Road Access, Off-Site Road Improvements, and On-Site Road Circulation

Regulations and regulatory staff influences will be imposed on the property layout by dictating access locations to the frontage road and connectivity to adjacent parcels. It is important to understand these issues prior to proceeding with conceptual land plans. The access location to the frontage road will be based on sight distance and safe turning movements. The location of the access may not be consistent with the development program for the property or be the best "marketing window." The condition of the frontage road may not be consistent with current standards, and improvements may be required. Know the requirements for off-site road improvements to determine costs and the impact on the development potential (Figure 4-2).

The on-site road circulation pattern is guided by the access location and connectivity that is beyond the site constraints.

Figure 4-2. Off-site road improvements are an important consideration by every level of government and should be investigated early in the due diligence period. The cost of off-site improvements may be significant and permitting time frames may affect the overall project schedule.
Source: David E. Johnson.

Land Development Regulations

In addition to the zoning map and regulations, the land development document enforced at the local level provides a comprehensive set of standards and criteria governing project design. It is important to obtain a copy of this manual to understand the development attitude of the community. In many cases, local municipalities have not modified their criteria to correspond with new sustainable planning approaches. Investigate the process in which standards can be changed to match the design intent of the project. Design waivers are not uncommon, and the town engineer may have the authority to revise standards for specific cases.

The land development manual addresses the technical criteria needed to design a subdivision or site plan based on the uses permitted in the jurisdiction. The standards are imposed for public rights-of-way and for private site development. All infrastructure design elements of a project are outlined in this document, and its approach to development has a direct relationship on the cost of development. As an example, this document will provide road widths, number of parking spaces required for retail commercial buildings, streetlight standards, and drainage inlet designs to performance bonding requirements. Every provision in the land development ordinance will have design and cost ramifications.

Many land development ordinances also refer to other documents for reference. Other agencies or departments within the municipality may be responsible for certain elements of the infrastructure design, construction, and maintenance. The most common is water and sewer. If there is a separate manual for water and sewer facilities, obtain a copy of this document as well.

In addition to infrastructure design standards and criteria, the land development regulations may address the regulatory review process. This process may consist of applications and forms to be completed with each plan submission. A timetable may also be included for government to act on a submission. A specific timeline for review and approval of land development projects may be included, and it is important to understand this procedure.

The land development ordinance is an important reference throughout the project development process, construction, and facility turnover. Know and understand this document because the provisions outlined will have a direct effect on the schedule, design, and cost of every type of land development project.

Stormwater Management Facility Requirements

Stormwater management (SWM) facilities control the quantity of stormwater runoff leaving the property. To summarize, this requirement states that the quantity of runoff after development must not exceed the quantity of runoff leaving the site before development. Knowledge of the technical approach to accomplish this task is not necessary, but you need to understand the local municipal regulations. These regulations include the design preferences, operation requirements, bonding, as-built requirements, ownership, and maintenance provisions. SWM facilities use land suitable for buildings, among other things, to control the quantity of stormwater runoff. Thus, the net buildable area will be reduced based on SWM requirements. Some provisions of the local SWM regulations may include fencing, landscaping, maintenance access drives, location, and turnover procedures.

Best Management Practices

As SWM facilities control the quantity of stormwater runoff, best management practices (BMP) control the quality of the stormwater runoff prior to leaving the site. Again, there are numerous ways to achieve the water quality standards, but local regulations will outline municipal preferences, construction standards, operation, ownership, and maintenance requirements. BMPs will require land for installation, and the cost of installation may be high. More important, know their regulatory approach to construction inspection, as-builts, bonding, maintenance provisions, and the turnover process.

Regional, State, and Federal Regulations

Compliance with local regulations does not necessarily ensure approval. Many local regulations conflict with regional, state, and federal regulations. Other agency regulations may take precedence over local standards and/or regulations. During the due diligence period, the regulatory map and site characteristic research should indicate which agency may have jurisdiction in permitting the project. At each level of government, more fees and review time will be required to obtain approvals. In most cases, the permitting process for regional, state, and federal agencies is not contingent on local approvals. But local approval is contingent on obtaining *all* permits.

It is the responsibility of the owner to obtain all permits that may be associated with the development of a site. As an example, the applicant may have to obtain a letter from another agency that simply states "no interest" in the project. Do not misjudge the importance of knowing the regulations and timeline of the agencies—at least know their overall regulatory charge and permitting timeline.

Of the following agencies, an applicant should at least expect review and permitting of the project by the soil conservation district, state EPA, and the USACE. Time is of the essence, so learn how each agency will affect your project schedule.

- Regional agencies: soil conservation district, watershed district, coastal agency, and transportation district.
- State government: Department of Transportation and EPA.
- Federal government: US Army Corps of Engineers, EPA, and Department of Housing and Urban Development.

Community Hot Buttons

Local government's unwritten regulation is the community hot button.

Entitlement, Approval, and Permitting Processes

The most important section of the regulatory analysis is the understanding of the plan review and approval process required by local government. Each local municipality has their own unique approach to assessing development applications. Most developers will indicate that a project schedule and the success of the project are predetermined during the regulatory review process. Time is money and delay during the review process is not uncommon, which only adds to the cost of developing land.

The zoning entitlement process is typically a different timeline than the approval of project plans based on approved zoning. The rezoning process may take longer and may have a defined set of dates in which milestones must be achieved, that is, community meetings, informal staff consultations, and so forth. This schedule cannot be modified because it is predetermined by the politicians at the beginning of each calendar year. The rezoning timeline will change by one month later if there is a problem with the rezoning application. Rezoning property is high risk with a potential high reward. Soft costs required to rezone property can be significant without assurances of obtaining approval. Take advantage of any informal meetings with politicians, neighbors, and staff to understand their points of view.

This section should also include a contact list for the stakeholders involved in the rezoning process. The review and approval of development plans may not be as stringent or legalized as the rezoning process. Obtain applications, fee schedules, and an outline of the process that may indicate a specific submission date before the planning board will act on the application. Alternately, the municipality may delegate plan review approval to the staff, and a time frame is provided in which the staff must act on the application. In either case, the timeline may be expedited;

include this option in the report. Provide the local regulatory timeline, steps for approval, and the contact list. The contact list may be different from those involved in rezoning cases.

Besides the written procedures, find out about board traditions and any "unwritten" nuances exercised during the regulatory process. Summarize the findings because "this is the way we do things" will directly affect the project schedule and possibly add cost to the project. Remember the developer's motto: Be on time and within budget. Following the regulatory plan review process exactly without alteration and with quality plans will minimize the time government takes to issue permits and approvals. However, in a no-growth environment, expect the worst. As an example, one town had an unwritten rule that the meeting would start at 7 p.m., but they would not start or accept any other application after 9 p.m. It became well known that the politicians and staff analyzed all aspects of the development in great detail, which limited the number of projects heard each meeting. It was not uncommon for a project to be included on the agenda but not heard for another six months.

Neighboring Property

Property adjacent to or near the PIQ will have an impact on the marketability and design approach. It is important to spend time on field reconnaissance of the market area. Location has always been a critical component of the decision tree, but adjacent uses could be problematic, especially for infill sites. Topography challenges could be a benefit to minimize less than desirable off-site improvements. The following lists the site analysis tasks necessary to determine the impact neighboring properties may have on the PIQ.

Existing and Proposed Uses

Start with a recent aerial photo to understand the existing uses of land adjacent to the PIQ. During the site visit, investigate all adjacent parcels for existing development patterns. Understand the relationship of the existing uses and development pattern and illustrate how the existing developments may or may not impact the PIQ's development approach. Take photos and inventory each photo for future reference. Find out if there is an association or a property management company involved with any of the existing projects.

Research the land-use map, master plan, and zoning map for the surrounding property, as well as for the property under consideration for purchase. It is important to understand the development potential of the adjacent property and how it may impact the development plans for the PIQ. It is conceivable that the adjacent property should also be purchased for subsequent phases of the project. The adjacent property may have attributes that would significantly increase the market appeal of the PIQ. Each parcel would be a puzzle piece in the development pattern, and how the property fits together for a comprehensive development pattern is advantageous for the community and developer.

Positive and Negative Surrounding Property Analysis

Prepare a detailed list of the positive and negative impacts surrounding properties may have on the development pattern of the PIQ. Consider technical, marketing, and political aspects of each property and describe how to overcome the negative and how to best utilize the positive aspects of the development, for example,

- Technical: The adjacent property may have an endangered species on the property requiring a buffer that may extend into the PIQ.
- Marketing: The proposed development may be consistent with the adjacent existing projects and both project designs must include road and pedestrian connectivity.
- Political: The PIQ may be adjacent to a 4,000 unit active adult community, and the intended use is first-time home buyer product.

Neighborhood Reaction to Development Plans

The opinions of citizens speaking out at public hearings will have a bearing on Planning Board or Council decisions. Addressing citizen input prior to a public forum is always the preferred plan for the applicant and public officials. A citizen public relations program could include

- Seeking neighborhood and community input on project plans and goals,
- Creating guidebooks on development costs and regulatory processing costs,
- Outlining citizen involvement in the process and define their roles,
- Preparing educational material on the public approval process,
- Educating newly appointed board members on the process,

- Developing an effective and mediation or arbitration system, and
- Increasing public awareness.

Political Assessment

It is important to know the political climate of a community. Is it a no-growth community or a community open to growth and new growth trends? With each local approval/permit requirement, the more influence the politicians have over the property and its use. Become familiar with

- Local politicians who would support your project,
- Staff who process projects and their influence,
- Regulatory permitting processes,
- Surrounding neighborhoods and if their leaders will oppose the project,
- Community hot buttons for inclusion in the project design,
- Experienced land use/zoning attorneys,
- Master plan for the area,
- Growth management plans,
- Subdivision approval process and the time required "to be in the ground," and
- Future road patterns.

Noise Problems

The most obvious noise producer is from departing and arriving airline flights. Check existing flight patterns and determine the possibility of future runway expansion. Additional runways and increased flights could alter flight patterns, which could adversely affect the PIQ.

In addition, traffic on interstate highways or arterial roads generate significant road noise, which travels farther than expected and is much more constant than airline noise. High speed and traffic counts have a direct bearing on the noise level. Designers understanding a noise problem could find cost-effective ways to overcome or significantly reduce the problem, such as raised earthen berms along the perimeter of the site with landscaping, window glazing, or leaving natural vegetation in selected areas.

As land values continue to escalate in price, properties located near industrial or manufacturing sites may become more affordable for residential housing. Of course, the use of the existing facilities will have a significant bearing on the viability of the project and market acceptance.

Views from the Site, Including Unsightly Views

Walk the property line and take pictures of the line of sight from the PIQ to adjacent properties. If there are unsightly views, makes notes on how to minimize the views. A few possible ways are to (1) provide a natural green buffer if the understory is thick, (2) prepare a land plan so the proposed lots are not backing directly along the line of sight, (3) design a landscaping mound with fencing, or (4) buy the property and include the land area in the project. Beneficial views of adjacent properties should be featured in the land plan.

Archeological or Historical Significance of the Site

Without a doubt, an archeological find could render the PIQ useless. The importance of undertaking an extensive property research for artifacts and historical elements is tantamount to a successful due diligence review. It is possible to set aside land for the discovery, but the marketability and public access to the site may not be worth the cost and time, time being the most important component. If the document search uncovers archeological or historical facts on or near the site, an assessment must be made on the viability of the site with a site constraint that cannot be altered.

Boundary Problems (Gores, Gaps, and Overlaps), Easements, and Rights-Of-Way

Site inspection should be undertaken after the boundary and topography base map has been prepared for the PIQ. There are four important property encumbrances that may appear on the base map and be highlighted in the title report.

1. Boundary line gores and gaps occur when the property lines between neighboring properties do not match and unclaimed property remains between the two properties. Assess the importance of the unclaimed property to the PIQ and ultimate subdivision plan. If it is of no consequence to the project, maintain the property line of the PIQ and allow the adjacent property owner to keep the unclaimed land.
2. Overlapping adjoining property lines are more of an issue, because each property owner owns a portion of the other's property. Local government will request property owners resolve boundary

disputes before moving forward with the subdivision plan. Overlaps tend to be small in area, but the property line configuration will determine if the extent or length of the overlap will be problematic for the project. Resolving this issue early is to the benefit of the project. If the land area and location has little to no effect on the PIQ, work with adjacent property owners and deed the areas to them. Time is more important than the value of the land contained in a gap or overlap.

3. Easements are typical for gas, water, sewer, and/or electric lines. It is common that utilities crossing a vacant parcel would be located underground except for the electric lines. Each utility entity has restrictions on the use of easements, and this must be incorporated into the project plans. Most utilities will remain in place owing to the cost of relocation.
4. Rights-of-way would either be set aside for public roads or for high-tension power lines. In some cases, the property line along the road frontage goes to the middle of the road, and government will request land to make the right-of-way of the frontage road consistent with regulations.

It is also possible that the adjacent property has a road right-of-way because of a commitment to a future road plan that may impact the PIQ.

Adjacent Property Impacts

If the PIQ is an infill site, the restrictions or influence of the existing infrastructure from neighboring properties will affect the subdivision layout. If the PIQ is located on the fringes of infrastructure, the subdivision plan and development program may have a direct impact on the neighboring properties, especially when connectivity among properties is promoted by local government. The extension of utilities and roads from the proposed subdivision to adjacent properties is a common requirement and must be duly considered during the layout phase of the project.

Environmental issues on an adjacent property may adversely affect the PIQ. No longer can you design in a bubble. You must assess the impact the proposed project has on the community as a whole and on adjacent properties, as well as their impact on the proposed project.

It is always a good practice to consider adjacent vacant property for expansion of the proposed project or for a separate project.

Site Characteristics

A comprehensive evaluation of a property's development potential should be undertaken early in the due diligence period (Figure 4-3).

In many markets, the availability of land suitable for residential development is reduced, and the parcels that remain have site constraints that will either increase costs or reduce the lot yield. Depending on the magnitude of the project and the environmental sensitivity of the area, developers may be required to submit an environmental impact statement (EIS). An EIS is a lengthy, time consuming, and costly report that is used to identify environmental issues that should be addressed during the initial phase of design.

Figure 4-3. Walk the property, including the approximate property lines. This on-site exploration provides you with an excellent idea of any issues that could affect costs and schedule. The intent of the site investigation is to discover problems that do not show up on any other maps.
Source: David E. Johnson.

For most residential projects, there are four primary site characteristics or constraints that will either increase project costs or decrease the number of lots, as discussed following.

Environmental Constraints

A property's natural characteristics should be used as an asset during the land planning process. With this as a goal, the existing environmental constraints imposed by all levels of government may affect the lot yield.

Specimen trees would be a definite asset for the project. The base survey should include locating all trees more than 24 in. in diameter and concentrate the survey in the areas to be developed. Locating the specimen trees will allow the design engineer a constraint that should be addressed during the layout design phase by limiting grade changes, highlighting selective clearing limits, and proving protective measures, that is, retaining walls and/or fencing.

The value added to a project far outweighs the replanting of new trees. Walking the property to identify specimen trees should be a standard practice, followed up by a field survey locating each specimen tree on the base map.

Wetland areas may reduce the buildable area only if isolated wetland areas can be permitted and filled or a wetland bank is available locally or statewide. In some locations, local municipalities have imposed a buffer beyond the wetland boundary. This wetland buffer could vary from 25 to 100 ft in width.

Even without a buffer, the design engineer should limit severe grade changes close to the wetlands' edge. Keeping construction traffic from the wetland boundary should be managed closely by the developer. Granted, not all wetlands are a marketing positive, and many jurisdictions are not considering wetland areas as a way to comply with open space requirements.

Buffers and conservation areas are imposed by local governments to lessen the impact on adjacent properties because of incompatible land uses, for retention of existing vegetation, or for environmental protection of lakes and streams. Conservation areas are left natural and restricted from many activities.

Some communities have imposed buffer regulations without much flexibility. Carving out buildable area for buffers and or conservation easements is not always a positive for the end user—the homeowner. This is especially true for areas with poor quality trees or old timber.

Stream corridors and jurisdictional streams not only restrict development options, but road crossings may be limited because of regulations, resulting in an increase in costs. Jurisdictional streams are delineated on soil maps and most GIS mapping. Jurisdictional streams are regulated by state and federal governments. The US Army Corps of Engineers may be the permitting agency, and this level of regulatory scrutiny becomes the critical path in the approval process.

Compliance with open space requirements, pedestrian connectivity such as greenways, and required passive or active recreation amenities will all add to the cost of development and further reduce buildable area. Of course, if used properly, these elements may increase home values and become a marketing positive for the project. A case can be made that these elements are not a detriment but should be considered an asset. However, long-term maintenance, repair, and replacement costs are typically the responsibility of a homeowners' association (HOA), which is rarely considered at the outset of the project.

Floodplains, ponds, lakes, and watershed protection areas may further reduce land area available for development purposes. Many properties used in the past for agricultural or livestock will have a farm pond on site. The pond should be assessed for structural integrity and whether the pond will be reclassified as a dam during the regulatory process. The pond should be assessed as a potential BMP and or stormwater management basin facility.

Always complete a document search for the latest floodplain map issued by FEMA. Floodplain mapping is updated and may have elevation information needed if filling the floodplain is essential to the project. A no-net increase of the floodplain elevations will be the benchmark at which the analysis will be reviewed and approved. The floodplain engineering analysis is costly, but the result may justify the investment. Local government should also be contacted to determine if the local jurisdiction has a floodplain administrator and whether separate regulations are imposed at the local level. Adjacent or nearby projects within the same watershed should be reviewed for floodplain information, if available.

Existing site buildings, structures, or previous land uses typically trigger a Phase 1 environmental analysis. This analysis identifies potential environmental or site issues requiring mitigation or cleanup. Groundwater contamination from any previous source on the site will definitely increase the cost of having an approved project. Time and money will be significant if groundwater is expected or discovered during testing.

Property Soil Types, Characteristics, and Topography

During the due diligence period, it is essential to best identify the different property soil types and their characteristics to determine the limitations soils will have on the cost effectiveness of the project. Combining soil information with the site topography will provide an excellent illustration of the site constraints directly affecting cost-effective land development activities.

The Soil Conservation Service, now known as the Natural Resources Conservation Service (NRCS), is an agency of the US Department of Agriculture and produces soil maps for the entire country. The maps delineate soil types on properties and list the characteristics of each. This information should be used for a quick planning assessment to determine the best soil testing program for the property. Concentrate on the upland areas with marginal soils and visible rock outcrops. Using resources in well-defined areas will allow a thorough investigation of those areas to ascertain the cost and development potential.

No specific soil testing program fits all sites. Some municipalities require a specific number of borings per acre. This might be a start in developing a soil testing program, but hiring a local soil material testing firm results in the best course of action.

Soil boring rigs are used for deep testing and require clearing access and setup areas. Soil borings provide a slice of information and are necessary in specific conditions and soil characteristics. Hand augers can be used to identify areas for further testing and are best used to identify wetland boundaries.

A test pit program using a backhoe is the best solution for wooded residential sites. This equipment can maneuver within the site and excavate a test pit to a depth of 12 to 16 ft. However, safety precautions must be implemented in accordance with Occupational Safety and Health Administration (OSHA) requirements when excavating at any depth. The type of soil dictates the safety measures to be taken by the excavator. Test pits provide a quick and cost-effective way for on-site soils to be clearly identified and mapped. This approach is best for finding the depth of unsuitable soil, type of soil, and depth of rock. A test pit is also best for soil mottling to determine height of the seasonal high-water table, finding the groundwater depth, and recording the thickness of the topsoil layer. For most sites suitable for residential developments, a test pit program is all that is needed. Working with the soils material consultant to define a practical and cost-effective soil testing program will provide enough information to assess the development potential of the site.

Soil erosion and sedimentation control of a site before and after development conditions has become the norm. The sophistication of this protective measure is directly related to the characteristics of the site soils. The soil testing program identify those areas on site where special protective features should be implemented and maintained until the project has been completed. The initial and ongoing maintenance costs are included in a contractor's price *if* the design plans and or project specifications outline the contractor's responsibility to maintain such facilities.

The existing topography shown on the survey base maps should be assessed by the design engineer before proceeding with the residential road and lot layout. Minimizing cuts and fills while striving to situate the home above the road should be a top priority. The existing topography may illustrate extremely steep slopes or relatively flat terrain, which all present a challenge to the designer. Many jurisdictions require road connectivity to adjacent vacant properties. Attention should be made to locate the road connection at the best possible location for the PIQ and any adjacent vacant land. The topography and property lines should dictate the connectivity.

Residential zoned land with topography and soil challenges may require the use of retaining walls and an increase in earthwork operation costs.

Topography shows natural drainage patterns and conveyance swales or channels. Most jurisdictions will not permit the redirecting of runoff from one drainage area to another. The civil engineer must prove the discharge from the improved site will not increase the flow before development and water quality standards are met. Natural drainage patterns and low points should be maintained and used to guide the land plan. As GIS becomes more reliable, a field boundary and topography survey are best for preliminary and final design.

Utilities

Water and sewer availability and plant capacities have and always will be essential to a profitable and cost-effective residential project. After the Great Recession of 2008 and during the prolonged housing recovery, many municipalities did not plan for redevelopment projects with land-use changes, mixed-use projects, and higher density smaller lot residential projects. The time required to obtain local utility approval and state permits has become lengthy, and in some cases, costly. During due diligence, the technical aspects of providing water and sewer service to the PIQ need to be assessed and regulatory fees outlined in detail. The only way to complete the utility due diligence section is to meet with the utility departments. Construction plans of existing water and sewer lines are not posted online and must be reviewed at the utility office.

Natural gas, electric, phone, and fiber optics are components of the infrastructure serving new residential projects. Each utility has its own timeline for design and installation and associated upfront fees. Prepare a comprehensive contact list for each company.

Water line extensions are not complicated and not considered an unusual cost consideration, as public water is essential. Water lines are relatively shallow and most all transmission lines are located in public road rights-of-way. The due diligence tasks for water include identifying the water distribution line pipe size and pipe material to which the project will connect. Request fire flow hydrant testing of the two closest fire hydrants. Flow and water pressure are key elements of the water design. Water availability and the quality of the final product are the responsibility of the local utility agency or municipal department. Future water supplies and distribution plans for a community's growth area may not be immediate, and the cost to implement a portion of the plan may be borne by the development community to ensure an approval of the project.

Other due diligence tasks include outlining the fees associated with each lot and the water rates charged to the end user. Also, ask the local utility about its policies concerning irrigation systems that would include backflow prevention devices and most likely a separate irrigation water meter. Before starting the land plan, obtain a copy of the standard water details for any deviation from the norm.

Sewer line extension requests typically generate a negotiated agreement between the developer and the municipality. This agreement is not prepared up front but rather later in the regulatory process because of changes in design and lot count. The agreement outlines the improvements necessary to serve the PIQ with sewer.

Using the site topography, due diligence tasks include determining if the property can be served by gravity sewer or whether a pump station with force main will be required. Using information provided by the local utility, determine if off-site sewer line improvements are necessary to serve the PIQ.

Due diligence tasks must include an analysis of the existing treatment plant to ensure adequate treatment plant capacity is available for the PIQ. With more stringent nutrient standards regulating wastewater treatment plant (WWTP) discharge, the plant capacity may be sufficient, but additional residential flow may not be accepted owing to the discharge nutrient limits imposed by USEPA and water quality protection agencies.

Because of funding limitations, municipalities find it difficult to properly size sewer trunk lines sufficient in capacity to manage future flows. Besides investigating the WWTP available capacity, due diligence efforts must also be directed to downstream sewer trunk line available capacity. Plant capacity may be available but the existing sewer trunk lines may not be able to handle additional flows, resulting in significant unanticipated costs.

Due diligence investigations should include identifying the regulatory fees assessed against the residential project and the time required to obtain approvals. It is recommended to ask for a "willingness to serve" letter from the local utility before a substantial investment is made buying the property and undertaking the design.

Government Staff Reviews and Comments

Each local government has its own ordinances, regulations, standards, criteria, and committee structure to control real estate development. A few of the local agencies are listed as follows:

Local planning board	Local zoning board
City council	Environmental commission
Shade tree commission	Town engineering/planning department staff
Town traffic department	Town building department
Town zoning department	Municipal utility authority
Fire department	Police department
Water company	Sewer agency
Water allocation agencies	School district
Park district	Improvement district
Drainage district	Lighting district
Air quality district	Floodplain district
Historical society	Landscape and architectural committee

Real estate developers may opt to establish an HOA for the ownership and maintenance of common open space, stormwater management facilities, and/or recreation amenities. HOA board members will also exercise their opinions on future phases of a project or on any other project proposed in the area.

Real estate developments are also reviewed at the county, regional, state, and federal levels, and by railroads and energy companies, that is, electrical and gas become involved in the review of a residential project if the project affects their existing infrastructure, easements, or land. President Ronald Reagan said, "No government ever voluntarily reduces itself in size. Government programs, once launched, never disappear. Actually, a government bureau is the nearest thing to eternal life we'll ever see on this earth."

Agencies and regulations do not disappear. In fact, many regulations and policies have not been updated to reflect current design and development trends. A perfect example would be the lack of design criteria for vertical mixed-use developments.

Dealing with the political environment and understanding the "no-growth" or "pro-development" culture of the community is essential to the success of a project. Each local approval has a combination of political and government staff opinions and recommendations. Because each municipality is different in its structure, it is essential to understand and appreciate the political influence on property. If the regulatory process includes a vote by a city council, real estate builder/developers must recognize that their personal viewpoints may override the intention of an ordinance or regulation and its use; this is called a community hot button, that is, an issue that has been elevated to their attention but not to the level of a regulation—at least not yet.

Residential builders/developers have more opportunities working with local politicians and staff than at other levels of government. Developers should become a part of the community and invest time and resources to assess practical and reasonable development patterns. They must become knowledgeable of the housing position of each politician. Would it not be advantageous to know which council member is pro-development? Knowing the political positions of council members is important but understanding the influences of the staff may be essential for obtaining an approval or permit.

Become an Active Member of the Community

There are basic, unwritten rules for residential builder/developers to follow for a positive regulatory review:

- Get to know local politicians and learn which may support a residential housing project.
- Meet with the project review staff and assess their views on residential developments.
- Read the ordinances, regulations, policies, and guidelines affecting residential projects.
- Determine which staff members hold the most influence and the most knowledge of the regulations.

Absolutely learn about the regulatory review and permitting process. Fees and the time required to "be in the ground" are affected by the government process. Meet with surrounding neighborhood leaders to gain their insights and resolve any issues prior to initiating the regulatory approval process. This is essential. Learn and understand the community hot buttons for incorporation into the project design program, if possible. Do not underestimate the importance of having a local experienced land-use/zoning attorney on the design team for guidance rather than for legal oversight. Read community documents, that is, master plans, growth management plans, zoning ordinances, land development guidelines, vision plans, transportation master plans, and other available materials.

Over the last 20 to 30 years, citizens have become more active in the regulatory process, and they are not shy about using their political influence to achieve results favorable to them. This will become even more apparent as development trends toward higher densities and mixed uses.

Citizens voice their opinions at public hearings, and it is in the best interest of the developer to include citizens in the review process. Seek neighborhood and community input on project plans and goals by setting up meetings ahead of the application process. Although this may be a duplicative effort once the application process is started, issues or concerns will definitely be discovered and addressed before government becomes involved.

Neighborhood meetings are an excellent opportunity to outline citizen involvement in the process and define their role. Developers risk the viability of the project by excluding citizen input on the design features of the homes or land plan. Most citizens do not have a clear understanding of the steps that must be taken to obtain government approvals and permits. It is prudent to prepare educational materials on the public approval process and distribute the documents before an application is made.

Create guidebooks on development costs and regulatory processing costs. Although costs are not considered a hardship, costs remain a deal breaker for any project. Most neighbors are more curious of the sales prices to be offered than the cost of development or cost of compliance with government.

Educating newly appointed board members on the development process may be warranted if the process is complex or lengthy to obtain approvals. Developing an effective mediation or arbitration system may also be needed if local government is receptive to the approach. There is an appeal process that may require legal action; however, most parties would agree that a negotiated settlement is better than a legal settlement.

Increase public awareness of the company and project with an active social media presence. Today's political arena requires developers to have a marketing, communications, or public relations professional on staff directing and managing the narrative throughout the regulatory process.

Government staff comments and influences may extend beyond their regulatory oversight. Some situations may dictate a boundary over which developers must escalate beyond the reviewer. This is considered a last resort but certainly justified if a delay is imminent or the cost to address a comment is unrealistic. Of course, the interaction between developers and reviewers should remain professional, but it is advantageous to know individuals' political sentiments toward residential projects.

Meet with the reviewer to discuss each comment and fully understand the intent of each comment. Written comments may not clearly define the problem or issue. In fact, the reviewer may be a new employee or may not even have technical background or experience in the design, permitting, and construction of a residential project. Typically, reviewers will follow their job descriptions and review projects based on their responsibilities:

- Planning and zoning department: This department is responsible for the zoning ordinance, subdivision regulations, and regulatory processes. Transportation planning, sidewalk plans, vision plans, and economic development may be the responsibility of the planning staff. Site plans, subdivision plans, and final plats are processed by the planning department.

- Engineering department: Engineering is responsible for the infrastructure on site and off site, including roads, drainage, stormwater management facilities, BMPs, and traffic concerns. The engineering department may also be responsible for site inspections and acceptance of the infrastructure for maintenance.
- Public works: This department is responsible for maintaining the rights-of-way, pavement conditions, solid waste pickup, and possibly recycling.
- Water and sewer departments: Water distribution, water pressure, water storage, and water treatment may be the responsibility of a department within local government or an outside separate agency. This responsibility would be the same for the sewer collection system, wastewater treatment, and effluent discharge. Residential growth has outpaced infrastructure improvements in many markets. Not only will utility staff consider the water and wastewater treatment capacities, it will analyze the distribution and collection systems from thc PIQ to the wastewater treatment plant. Contact the utility staff and ask for information concerning the system status from water treatment to the wastewater treatment plant and ultimate discharge to a receiving stream or river.
- Building department: The building department is typically understaffed, as most are fee based. Find out how long the review process takes. This staff is responsible for reviewing building plans for code compliance. Interpretation of code stipulations can be appealed to the code, if needed.
- Police department and emergency services: Staff will review the road layout of the subdivision and be more concerned about the response time to the project, especially if located on the outskirts of town. Street and house addressing will be important for identifying the proper location. Gang mailboxes are now required for residential subdivisions. Street addressing on the curb or home becomes much more important regardless of the technology available to find a home.
- Fire department: The fire department, and specifically the fire marshal, has become a noteworthy influencer in the review process. The fire marshal's review is based on a national code adopted locally. The National Fire Protection Association issues a code every three to five years with revisions taking up to two years for adoption into the code. This code is typically overlooked by developers but is weighted heavily by local politicians because the code was created

to protect lives and property. One of the most recent regulations affecting residential development is the requirement of having a second road access into the project if the project has more than 30 homes.

Government regulations can be interpreted, and intent can be debated. The government staff will either be supportive in a pro-growth environment or be unbending in a no-growth community. Developers and land development engineers should create and promote a professional working relationship with the review staff of all agencies. This relationship fosters quality projects for the builder/developer and the municipality.

Development Potential Summary

Technical site analysis work is important to assess the land for real estate development use. Developers of residential subdivisions typically strive for maximizing the number of lots. The lot layout and yield will reflect the technical site analysis conclusions and specific government regulations. The cost of land development activities will vary but not drastically. Developers should maintain a land development cost basis for use in determining the per lot cost using a conceptual land plan. Over the last several years, the cost of developing property has increased in some markets substantially.

Once a land plan and a conceptual cost estimate are prepared, these costs are tracked through the entire approval and permitting process. Ultimately, the conceptual cost estimate should be compared to the prices submitted by contractors.

Chapter 5

Financing

In structuring land development projects, you will want to investigate the many forms of financing mechanisms. Joint ventures, installment contracts, and more traditional financing opportunities are available to finance housing projects. Financial planning is an ongoing process that should match the company's commitment for growth. This chapter provides an overview of alternative financial approaches for funding small to midsize land development projects. The larger and more complex projects require a much more in-depth financial analysis than the basics outlined in this chapter.

Financing Techniques

Knowing your site and the ultimate use of the property is essential in preparing to obtain financing for the project. A well-conceived project plan, demonstrating excellent management skills, effective management of cash, and sufficient financial resources, are also indispensable for a successful project. All lending institutions will require you to present these elements of the project.

The purchase of raw land offers the greatest potential return on investment of any real estate venture, so the selection of an advantageous financing package is important to the operations and financial position of the project. Using any of the alternative financing techniques does not alleviate the need to fully understand the project's financial needs. Undeveloped and unapproved land does not produce income or cash flow; raw land generates only mortgage payments, interest costs, taxes, insurance, and other payments as dictated by the signed contract. Time is essential in converting raw land to finished lots. The cost of carrying the land until approved and improved may very well drain any profits generated from the initial investment. Obtaining permits to start constructing infrastructure takes time and money. Large developers have an advantage over local developers because of their financial strength and ability to weather the unexpected.

Financing land development projects takes many forms and the partnering of diverse interest groups. Typical financing techniques are described in this chapter for ease of reference. Conventional financial controls on land development projects may become more stringent and inflexible. Thus, developers should investigate alternative financing techniques for future projects.

Purchase Money Mortgage

For this mortgage, the purchaser obtains a legal title to the property. The seller takes a mortgage as security by receiving a down payment and in the contract indicates the periodic payments for the principal and interest. If the purchaser defaults, the seller takes back the property. The purchaser must ensure that he or she is free to quit the contract if the loan goes into default. Thus, for developing raw land, design a way out of a project and do not put design dollars into the project until all certainties have been proven. A prudent business approach includes contractual provisions that provide escape clauses in case all valued market estimates, opinions, and anticipated market conditions fail to materialize.

Options

For a negotiated price, the purchaser obtains exclusive rights for a fixed period of time to buy land. The seller retains full ownership until the purchaser exercises this option. If the purchaser experiences adverse market conditions, that purchaser can be released from the contract and suffers only the financial loss of the option price. This is an inexpensive way to control property over a period of time, and it allows time for favorable market conditions, rezoning, approvals, and so on.

Installment Contracts

The seller retains ownership of the property but prepares a negotiated down payment with the purchaser and lists subsequent payments of principal and interest. This approach is typical with landowners who can retain ownership and have no pressing need to sell the land, as sellers with a purchase money mortgage do.

Leasehold Arrangements and Lease Options

A purchaser makes a lease arrangement with a seller from whom an outright purchase is not possible.

Land Sale Lease Back

The landowner sells the property to an investor, who then leases the land back to the developer for eventual development.

Limited Partnership

The developer or general partner provides the management overhead and expertise to manage the partnership and its business activities. The investors provide the funding. The risk is spread to those involved in the limited partnership activities.

Syndications

Syndications raise capital resources through numerous investors. The syndicator manages and invests the cash raised. Many regulatory disclosures and requirements are imposed on syndications.

Joint Ventures

A joint venture is an excellent technique for combining the forces of the landowner with those of the land developer. The developer and landowner split a piece of the project, and the joint venture structure reduces the financial liability of the developer. The profits, losses, and liabilities can be allocated for each specific piece. The specifics of each joint venture agreement must be clearly defined in the contract agreement. Even for large companies, the joint venture financing approach is most often the best technique for developing land in almost every state in the country.

Each state interprets the laws governing joint ventures differently. A joint venture is an association of two or more parties carrying out a single activity for profit. It requires a common interest in that purpose, and the parties have the same right to direct and manage the conduct of each other in connection with the activity. Expecting profit and sharing management roles are essential elements of a joint venture. If the agreement does not include a definition of the sharing of any profits, the agreement would imply equal sharing among the joint venture partners.

The following checklist can guide the structuring of real estate joint venture agreements:

1. Start and term of joint venture,
2. Powers of joint venture group,
   - Hold title
   - Acquire, lease, or sell real property
   - Borrow or lend funds
   - Pledge assets
   - Enter into contracts
   - Hire employees
   - Retain consultants
   - Enhance joint venture business
3. Role of joint venture partners,
4. Indemnification clause,
5. Management control,
6. Contribution to venture group,
7. Fiscal year definition,
8. Accounting responsibilities,
9. Arbitration provisions,
10. Distribution of profits and losses,
11. Buyout provisions,
12. Transfer of interests,
13. Termination clauses, and
14. Execution.

Blanket Mortgage

A blanket mortgage covers more than one parcel of real estate and in general is associated with single-family housing projects. The lending institution releases individual parcels or lots from the lien on the property by the lender.

Other Financing Techniques

Short-term financing, interim financing, open-ended commitments, commitments with guarantees, and permanent take-out commitments, including financing options, are recognized financing instruments. These approaches can be discussed with accounting consultants or lending institution representatives for more detailed information. These approaches are reviewed for short-term and interim financing requests, whether the commitment is an open-ended commitment, commitment with guarantees from the developer, or a permanent take-out commitment that is converted at the completion of a project into a permanent mortgage loan.

Once the financing package or technique is considered for a single-family subdivision project, investigate what the lending institutions are expecting in terms of presentations by the developer. Developers need to understand the position of the lender. Simply, lenders are looking to be repaid and not to take back the property. Also, lenders need for developers to understand the bureaucracy and the time it takes to process a loan request. Lenders require capital investment in the project by the requestor.

Public–private partnerships provide a unique way for local governments and private developers to work together to design and build residential infrastructure. This association provides an avenue by which a community can experience sustainable growth while the risk of private investment is minimized. This is an opportunity for many communities to generate rooftops that ultimately spur the demand for retail, commercial, and business uses.

Choosing a Lender

It is just as important for the developer to qualify the lender as it is for the lender to qualify the developer. Learn what the lender expects in the loan submission package so the submission answers many of the lender's questions up front. This shows that the developer has done his or her homework in preparing a professional approach to the project while accommodating the needs of the lender. Preselecting the lender is important to limit the amount of time and effort in presenting the loan request. A blind mass distribution of the loan package to lenders gives the impression that the developer is shopping around—and maybe even that the developer is desperate for financial support. Therefore, preselect lenders and concentrate activities, energies, and cash on developing the proper approach for a select few lenders to ensure a higher rate of success in obtaining a loan commitment.

In selecting the lenders most likely to qualify a developer for a project, use the following guidelines:

1. Evaluate all the available financial sources, that is, a commercial bank, a real estate investment trust, joint venture possibilities, private investors, and so forth.

2. Ask for the lender's references.
3. Determine the amount of funds available or the amount that has been set aside for real estate lending.
4. Determine the maximum loan size and credit extended to one borrower and determine if the loan request is within those limits.
5. Determine the percentage of capital that can be invested in the real estate industry or real estate loans. What is the current investment limit?
6. Determine whether the lending institutions have any loan problems or are under review at the state or federal level.
7. Determine the internal policies, practices, and/or regulations for real estate lending.
8. Determine and review the current real estate loan portfolio and the list of clients.
9. Determine the lender's geographic capabilities and whether interstate lending is an option.
10. Obtain, if possiblc, a sample loan approval or submission package.
11. Determine the processing requirements of the loan package.

The best time to determine who is lending funds for land development and building activities is during the market research and site analysis stages. Visit competitive projects and look for jobsite signs indicating the financial institution partner for the project. Network at HOA meetings. Contact real estate agents and brokers and meet with the competitors to understand who is and who is not lending. Assess each lender's strengths and weaknesses. Understand the bureaucratic approach of each lender to a housing project. Assess the attitude of the institution toward builders and developers.

During the loan submission process, again evaluate the lender's current record in lending for residential developments. The substantial upfront costs for project planning, including consulting fees, application fees, and escrow fees, require that you understand the lender's policy on financing these costs. If the lender covers start-up costs, the amount of capital infusion by the borrower is less. Also, determine how overhead costs are included in the financing package. During the initial discussions with the lender, ask about the performance bond capabilities of that lender, and ask how the performance bonds can be included in the loan package. In real estate, the most important policy of the lending institution is how it determines that a performing loan has become a troubled loan; learn the ramifications of becoming a troubled loan. Know this policy before submitting the loan package.

Be prepared to submit a complete loan package once the lender has been selected. Remember, lenders look for professionalism when qualifying builder/developers. Lenders appreciate borrowers whose standards include proven business practices. Adhering to the lending institution's paperwork requirements for a relatively small project is just as significant as that required for a large project. To minimize the paperwork, develop efficient ways to keep the lending institution informed of the financial status of the job.

Submitting the Loan Package

Submitting a completed loan package demonstrates your ability to provide correct and complete information the first time. The lender's comfort level is directly associated with the credibility of your presentation at the very beginning. A loan submission package should contain the basic items listed next. The overall tone of the loan package should be positive and enthusiastic about the project.

Cover Letter

The first element of the loan package is the cover letter, which contains a brief description and location of the proposed project. Give a short summary of the financial request and highlights of the project. Be positive, to the point, and brief. The cover letter should be approximately one page in length and no longer than two pages. Determine the specific person who will review the initial loan submission package. The cover letter should be personalized, and the loan package should be submitted directly to that person. Call the reviewer to schedule follow-up meetings once the loan package has been submitted.

Loan Summary

A one-page summary of the loan request is important so that the lender can immediately recognize the limits of the request and how it conforms to the institution's rules, regulations, and lending policies. Provide a summary of the transaction parameters with an outline about collateral and the type of financing requested. Indicate the timing for the project and any other important information that the bank would need to act on the loan request.

Project Elements

This section also should be brief and to the point. List and explain the housing types to be constructed in the project. Define the target market and the price range of the product. Include a location map for the lender to assess the importance of the project location. Define the absorption rate and outline the reasons why this project will succeed.

Company History

Include a summary of the company's experience in the housing industry. Highlight experience with the proposed product type. The personal financial capabilities and credit history of the borrower and of the company must be outlined in the loan package. Additional information concerning any community activities, key staff members, and their credentials should be included in the request. List all the investors.

Marketing Strategy

Include an executive summary of the market study (as outlined in Chapter 3). The executive summary is a concise statement outlining the highlights and conclusions of the market research. This, of course, would include the geographic market area, demographics, building activity, trends in the marketplace, and projections supporting the financial request.

Project Team Members

List the consultants who have been retained to provide services for designing, planning, and permitting the project. The various disciplines should include the engineer, soils engineer, surveyor, environmentalist, planner, architect, marketing consultant, ad agency, public relations agency, and interior decorator. A brief list of these firms and key individuals assists the bank in understanding your commitment to the project by retaining well-qualified people to implement the project plans.

Graph of Competitive Product

Provide a brief and concise graph of the proposed product versus the product of the competition. Also include the summary of the site analysis. Show how the project will use existing infrastructure and be consistent with municipal master plans.

Target Market

Include a one-page summary of the target market. This summary should also be brief and concise. The description of the target market includes a listing of the age ranges, family composition, income levels, employment characteristics, lifestyle information, and financial capabilities of the market.

Product

Summarize in one page the product line proposed for the project. Include the product design, house layout, subdivision layout, density, unit size, and product specifications. Highlight certain features that show how the proposed housing product has a greater value than the competition. Emphasize the pricing and absorption levels proposed for the project.

Appraisal

Include the most recent appraisal of the property.

Sales and Marketing

Include a list of the top sales agents, their experience levels, credentials, and résumés. Outline the advertising and promotional program for the project. Include brochures and other material supporting the marketing approach.

Pro Forma

Appendixes 5-1 and 5-2 are project pro formas summarizing the cash flow requirements anticipated for the probable scenario and the most conservative scenario. Provide detailed cash flow scenarios that represent the most probable case and the most conservative and worst cases for the bank to analyze. These pro formas should be prepared using computers and available spreadsheet software specifically designed for builder/developers. They are not only beneficial for the presentation of a loan request, but they will provide a means of keeping track of the project progress during construction. Backup schedules are required to show how the costs were calculated for each expense line item. This detailed cost analysis approach is beneficial to the investor and/or lender for their objective review of the project's potential.

Corporate Reports

Include any reports generated by the company for similar types of projects. This may include weekly reporting of traffic and sales information. Also, construction schedules and accounting reports are necessary for the bank to review. The bank will determine whether the reports are sufficient in detail for proper monitoring of the project. Provide a methodology or a means of reporting the construction activities on a continuous basis to the lender. In essence, develop a management report that provides the sales, absorption, and actual cost versus budget analysis for the project. This financial report is the best way to keep the lender informed as the project progresses.

Financial Documents

Include any corporate or personal financial statements and any other bank financial references. List the contact person who would be involved in follow-up discussions and meetings.

Miscellaneous

Also include in the loan package a set of improvement plans prepared for the project. These plans include the final plat, engineering plans, landscaping plans, environmental reports, aerial photos, renderings of the product, floor plans of the product, photos of the site, photos of surrounding areas, and project schedules. The schedules include land development operations and the timetable to complete the first house.

Every company has strengths that should be presented with enthusiasm in the loan package. If possible, prepare your loan package documentation with professional flair using desktop publishing. An excellent visual presentation of the material is beneficial during the review process. A strong first impression adds credibility to the facts and figures that the bank will use in assessing and approving the loan request.

Appendix 5-1

MANAGEMENT PROFORMA - PROBABLE CASE

Sources of Cash		Year 1	Year 2	Year 3	Total
Number of Closings		6	36	2	44
Avg. Price $191,397		$1,148,382	$6,890,288	$382,794	$8,353,259
Sale of 25% Interest at $25,000 Per Share: 10 Shares		$250,000	$0	$0	$250,000
Total Cash		**$1,398,382**	**$6,890,288**	**$382,794**	**$8,603,259**
Uses of Cash	**Per Unit**				
Architect	$227	$2,000	$8,000	$0	$10,000
Engineer	$2,409	$51,500	$54,500	$0	$106,000
Land	$15,000	$45,000	$585,000	$30,000	$660,000
Land Development	$22,000	$545,000	$423,000	$0	$968,000
Offsite Improvements	$0	$0	$0	$0	$0
Home Construction Costs	$103,280	$1,134,939	$3,323,601	$85,775	$4,544,315
Closing Costs	$500	$3,000	$18,000	$1,000	$22,000
Warranty Service	$500	$500	$16,000	$5,500	$22,000
Project Overhead	$5,068	$68,605	$147,404	$7,000	$223,009
Real Estate Commissions	$3,781	$20,904	$137,804	$7,656	$166,364
Sales and Marketing	$3,837	$91,989	$76,832	$0	$168,821
General and Administrative	$5,659	$32,405	$206,708	$11,484	$250,597
Finance, Insurance, Tax,...	$5,926	$115,750	$140,000	$5,000	$260,750
Interest/ Bank Debt(10%)	$2,454	$34,300	$73,667	$0	$107,967
Operating Uses of Cash	$170,641	$2,145,892	$5,210,516	$153,415	$7,509,823
Cash Flow from Operations		($747,510)	$1,679,772	$229,379	$1,093,436
Loan Proceeds from Bank		$1,705,000	$3,747,670	$0	$5,452,670
Loan Repayments		($835,000)	$1,069,124,154	$0	($5,452,670)
Return of Principal		$0	($250,000)	$0	($250,000)
Net Cash Flow		$122,490	$559,772	$229,379	$911,641
Total Profit Distribution					
Investors 25%					**$227,910**
Company 75%					**$683,731**
Total Profit Distribution					**$911,641**
Outstanding Bank at Period		$870,000	$0	$0	
Outstanding Invest at Period		$250,000	$0	$0	
Cumulative Cash Balance		$122,490	$682,262	$911,641	

Appendix 5-2

Management Pro Forma: Conservative Case

Sources of Cash		Year 1	Year 2	Year 3	Total
Number of Closings		6	26	12	44
Average Price $185,000		$1,110,000	$4,810,000	$2,220,000	$8,140,000
Sale of 25% Interest at					
$25,000/Share: 10 Shares		$250,000	$0	$0	$250,000
Total Cash		$1,360,000	$4,810,000	$2,220,000	$8,390,000
Uses of Cash	Per Unit				
Architect	$227	$2,000	$8,000	$0	$10,000
Engineer	$2,409	$51,500	$54,500	$0	$106,000
Land	$15,000	$45,000	$585,000	$30,000	$660,000
Land Development	$22,000	$545,000	$423,000	$0	$968,000
Off-site Improvements	$0	$0	$0	$0	$0
Home Construction Costs	$103,280	$1,134,939	$3,323,601	$85,775	$4,544,315
Closing Costs	$500	$3,000	$18,000	$1,000	$22,000
Warranty Service	$500	$500	$16,000	$5,500	$22,000
Project Overhead	$5,068	$68,605	$147,404	$7,000	$223,009
Real Estate Commissions	$3,781	$20,904	$137,804	$7,656	$166,364
Sales and Marketing	$3,837	$91,989	$76,832	$0	$168,821
General and Administrative	$5,659	$32,405	$206,708	$11,484	$250,597
Finance, Attorney, Insurance, Tax, etc.	$5,926	$115,750	$140,000	$5,000	$260,750
Interest on Bank Debt(10%)	$2,454	$34,300	$73,667	$0	$107,967
					$7,509,823
Operating Uses of Cash	$170,641	$2,145,892	$5,210,516	$153,415	$7,509,823
Cash Flow From Operations		($785,892)	($400,516)	$2,066,585	$880,177
Loan Proceeds From Bank		$1,705,000	$2,747,670	$1,000,000	$5,452,670
Loan Repayments		($835,000)	($2,300,000)	$1,071,424,154	$1,068,289,154
Return of Principal		$0	$0	($250,000)	($250,000)
Net Cash Flow		$84,108	$47,154	$229,379	$360,641
Total Profit Distribution					
Investors 25%					**$90,160**
Company 75%					**$270,481**
Total Profit Distribution					**$360,641**
Outstanding Bank at Period		$870,000	$1,317,670	$0	
Outstanding Invest at Period		$250,000	$250,000	$0	
Cumulative Cash Balance		$84,108	$131,262	$360,641	

Chapter 6

Schedules

Developing raw land into finished lots is an art. It can be financially rewarding as long as the project stays on schedule and within budget. This chapter explains the various practical scheduling techniques that can be implemented for residential housing developments. The intent is not to present a sophisticated critical path method (CPM) or a sophisticated program evaluation and review technique (PERT) scheduling method but to illustrate proven scheduling techniques that can be used to track work production.

Project Activity Schedule

Schedules are valuable when used on a regular basis. Appendix 6-1 is a comprehensive activity schedule for tracking a project from land acquisition through the final transfer of land improvements to local government. This activity list presents a time frame for each housing activity; note that many activities run parallel with each other. The time frames in Appendix 6-1 are representative of a typical housing project schedule.

To expedite any project, select activities for which the timeframe can be modified. As an example, consultants require a certain amount of time to prepare the designs. However, in this area, the timeframe can be accelerated. Consultants make money by completing projects in a timely manner to have staff available for the next project. Ask them to expedite your project. Conversely, once your project has been submitted to the regulatory agencies, your ability to accelerate the permit process becomes very difficult, if not impossible. Successful projects must be managed, and the developer personally should manage the schedule. The regulatory process and public hearings can be managed successfully by keeping track of details.

The timeline in Appendix 6-1 also can be used during contract negotiations and to determine the contractual time limits for the project. The first phase, the conceptual design, establishes the project design parameters. Because the conceptual design dictates project costs, it should be approved by company staff members before the expenditure of funds for the preliminary and final design phases.

When the improvement design has been completed, initiate the regulatory process at all appropriate levels of government that are required to review the improvement plans. When government agencies provide a review letter seeking additional information or requesting revisions, the plans must be resubmitted as quickly as possible. The ability to transform regulatory issues into project objectives is a critical element in responding to regulatory recommendations. Keep the pressure on government to efficiently review, approve, and permit the housing project. If your response is not completed expeditiously, your project may receive a lower priority for review, and subsequent resubmissions, if required, may not receive proper governmental response. Therefore, it is incumbent on the developer to ensure that the consulting team responds effectively, efficiently, and quickly to governmental problems and issues. Consultants should design the plans in accordance with regulatory standards to avoid resubmissions.

The last section of the comprehensive project activity schedule (Appendix 6-1) refers to the municipal acceptance of infrastructure (Figure 6-1). The transfer of public rights-of-way or facility improvements has become a difficult, time-consuming, and costly process. The key to finishing a housing project and its land development improvements is to turn over the infrastructure to the government entity as quickly as possible.

Companies that effectively manage quality control and contractor punch lists stand a better chance of passing regulatory inspections and turning over public improvements promptly. Therefore, inspect construction regularly to ensure that quality is maintained. When quality control is apparent, government officials are less likely to question the quality, and punch list items are minimized. Once land development contractors have left the jobsite, scheduling their return can be difficult to take care of any punch list items that need to be addressed. Thus, to ensure a timely performance bond release, the developer should hold a land development retainer to ensure the punch list items are completed promptly by the contractor.

Developing a single-family subdivision is similar to producing a product at a manufacturing plant. The raw land enters the plant door; goes

Figure 6-1. Infrastructure costs. Material and labor costs for infrastructure elements continue to increase. Drainage features are a significant percentage of the total land development cost of a project. Source: David E. Johnson.

through design, costing, bidding, permitting, and construction phases before being approved to leave the plant; and is then finished and wrapped with the keys handed to customers. A manufacturing process of delivering houses on a continuous basis is predicated on the fact that gaps or lapses in the production schedule do not delay production. Production gaps affect a company's financial stability, employment base, and reputation. Management tools tracking costs, schedules, reports, communication, and plans are essential for being on time and within budget. Field management can be managed through cloud-based software packages with immediate access for anyone on the project team from anywhere on the project site. The short list of software packages worth reviewing are Procore, Buildertrend, Bluebeam, PlanGrid, and e-Builder. All these management tools and others provide the resources to link field management with the office, subcontractors, and local government.

Master Schedules

Appendix 6-2 is a project master schedule that provides a scheduling overview for multiple projects. This approach highlights key elements of the project completion schedule from the land planning phase through to the certificate of occupancy. The first certificate of occupancy and the first closing inject cash flow back into the company. Monitor the progress of each project within the inventory so that closings can be predicted and production gaps minimized.

Appendix 6-3 illustrates an individual project master schedule, which presents projected due dates for each major activity line item in the development process. It also shows the actual dates of completion for each line item. A quick review of the schedule will identify the projects ahead of schedule as well as those projects in serious trouble. At the land acquisition stage, this scheduling methodology is very useful in predicting the first closing in the project. After completing the project master schedule, work backward to establish a start date, and use this information at the land purchase negotiation table. The project master schedule also tracks submission due dates for regulatory agencies; these dates are important because most agencies require a certain amount of time to review plans before official action is taken. If these submission dates are not met by the consulting team, delays in the project planning process will occur. By missing a scheduled submission date, the project could be delayed two to four weeks. A delay caused by a missed scheduled due date may affect the project schedule even more if seasonal concerns (e.g., winter, rainy season, and temperatures) also influence operations.

Land Development Schedules

The land development project schedule in Appendix 6-4 highlights the major activities of a housing project and how long it takes to complete them. This format can be adapted to any project consistent with the company's business plans. All major land development activities are listed on this chart, as well as actual progress. Track the various activities on a daily, weekly, or monthly basis. Highlight work stoppages or delays and indicate whether liquidated damages are to be charged against any subcontractor. Highlight milestones for each activity on a weekly basis, including the percent completed. This management tool helps in monitoring work progress and in approving and processing invoices.

Illustrate on the schedule any problem areas that must be resolved by a business, regulatory, or permitting decision. As work progresses in the field, track these activities in an illustrative way (see Chapter 19) and transfer the information to the land development project schedule. On the basis of the site analysis phase, prepare the land development project schedule using the industry production ranges presented in Table 6-1.

The land development project schedule can be prepared based on a conceptual land plan. This timetable, when combined with a house timeline, provides an estimate for the closing of the first house. Depending on the time of year for construction, factor in slack days for weather, holidays, hunting season, and so on. Labor production rates fluctuate with the crew's skill level for each land development activity. Alter these ranges to fit site conditions, quality of workmanship, and quality of the available labor force.

With personal computers, it is easier to develop schedules and reports to assist in the overall management of a housing project. Many software packages are available for scheduling housing construction. Because the land development tasks vary considerably for each project, many developers continue to prepare land development schedules by hand. Manual scheduling requires daily hands-on knowledge of the development work; delegation of this responsibility to an on-site superintendent should relieve the owner from updating the schedule on a daily basis. Many land developers equip their field superintendent with a laptop computer with direct access to the main office. The schedule is updated daily using Microsoft Project or similar programs providing general scheduling methods. A daily review is still necessary, however, because land development contractors can spend tremendous sums of money in a short period of time.

Delays can be expected during the normal progress of a land development project. Many problems that occur on site are not described on a management report until after the fact. The land development schedule provides a means to track project status versus value received and to follow problems daily. The schedule consolidates the progress of various subcontractors, rain days, or problem delays on one management form.

The land development schedule also provides a historical record of the project, which can be useful to improve the management of the next land development project. In addition, this historical information provides a comparison of work productivity of the subcontractors. Their ability to get the job done on time and according to the contract agreement is an excellent reference for future work. This schedule should incorporate

Table 6-1. Daily Production Rates by Task.

Task	Rate	Factors
Clearing	0.5 to 1.5 ac/day	Depends on the maturity of the site, the species of tree, and the density of growth.
Excavation	1,000 to 2,000 ft^3/day	Varies by soil type and design. Based on using a front-end loader and trucks to move the material. Rate can be doubled if the design accommodates the use of pans instead.
Sewer	300 to 400 ft/day	Based on installing an 8 in. sewer line at a depth less than 8 ft. Contractors specializing in underground utility installations can offer lower unit prices and increase the rate to more than 900 ft/day.
Water	500 to 1,000 ft/day	Varies by the design and soil type. Rate is higher than for other underground utilities because the depth of construction is constant and greater installation distances can be achieved without interruptions.
Drainage	200 to 600 ft/day	Rate for pipe sizes up to 30 in. As pipe size and the number of junction points increase, the daily production rates decrease. The type of pipe material also affects the rate.
Curb	700 to 1,000 ft/day	Based on straight-line concrete barrier curb installation. For Belgium block or mountable curb installations, the rate increases with the experience level of the crews.
Paving	3,000 to 4,500 sq. yd/day	Based on residential roadway specification. Contractors specializing in residential paving will have experienced labor and equipment available to pave up to 10,000 sq. yd/day.

Note: ac = acre, ft = foot or feet, in. = inches, ft^3 = cubic feet, sq. yd = square yard or yards.

the contractors' schedules and be agreed on before the mobilization of equipment. If properly used during land development, developers can maintain control of their jobs, minimize delays, and reduce extras, with the result that the land development work will be completed on time and within budget.

Housing Inventory Schedule

In project planning for production housing, a rule of thumb is to allow land development construction activities to precede housing activities by approximately three months. This permits underground utility work, road construction, and electricity and telephone installations to be completed for each phase of construction (Figures 6-2 and 6-3).

Proceeding with land development before housing construction will eliminate any confusion on site and any conflicts among housing subcontractors, material suppliers, and land development contractors. It is

Figure 6-2. Utilities. Contact all utility companies early in the project planning phase and request utility layout plans. Proper scheduling of utility companies for installation is critical to meeting the overall project schedule.
Source: David E. Johnson.

Figure 6-3. Scheduling. Most residential builders will seek opportunities to proceed with housing construction while land development activities are being completed. Proper phasing will separate housing and land development contractors within the project site.
Source: David E. Johnson.

always preferable to start housing construction after the site is stabilized. Housing subcontractors then have easy access to each home under construction. In addition, the advance site preparation enables buyers and potential buyers to inspect the homes under construction without crossing the path of land development equipment. Access to the homes under construction and site inspections of available lots encourage referrals. Schedule land development activities to complement the housing production schedule for each housing delivery.

Appendix 6-5 tracks the housing inventory by project. This housing production schedule follows each housing sequence under construction. List either the lot and block, or track each house by a sequence number, in each phase of the project. This schedule records the original start date and the actual/forecasted date for each unit. It also lists, by month, the completions, closings, and inventory units, which should be limited to models. This report also tracks sales orders and sales backlog with the starts and closings. By completing this report on a month-to-month basis, a company can assess the need to start purchasing new raw land or approved housing projects to supplement any shortfalls in housing production.

Appendix 6-1

COMPREHENSIVE PROJECT ACTIVITY SCHEDULE

DAY	ACTIVITY
Land Acquisition	
000–030	Project Evaluation
030–090	Access Market, Competition, Labor Force, etc.
000–030	Regulatory Analysis
030–090	Site Analysis, Technical Review of Site Information
060–120	Product Development and Density Analysis
120–135	Prepare a Proposed Land Development Budget
135–135	Prepare Project Proforma
000–180	Negotiation Strategy, Prepare Contract for Land Purchase
Conceptual Design	
150–180	Select Design Team Members
030–120	Indepth Regulatory Review with Design Team
150–180	Finalize Design Development Program for the Project
180–210	Prepare Conceptual Land Plans for Review
180–210	Develop Preliminary Housing Design Program
180–270	Develop Project Theme, Market and Sales Programs
220–220	Select Land Plan and Finalize Housing Concepts
210–240	Complete Conceptual Engineering Design
210–240	Complete Design Development Architectural Plans
240–270	Prepare Conceptual Cost Estimate and Budgets
270–270	Complete Project Proforma
240–270	Establish Project Goals, Schedule and Opening Date
240–270	Review Construction Specifications for Project
280–Completion	Monitor Overall Project Schedule

Preliminary and Final Design

240–270	Negotiate Professional Services Contracts
270–390	Complete Final Engineering Design Plans
270–390	Complete Final Architectural Design Plans
330–390	Complete Landscaping Design Plans
360–390	Prepare Preliminary Cost Estimate and Budgets
390–390	Prepare Project Proforma

Regulatory Process

240–390	Schedule Planning Sessions for Regulatory Strategy
390–480	Municipal and County Planning Boards
390–480	Municipal Planner and Engineering Review Meetings
390–450	Water and Sewer Municipal Review
390–450	Soil Conservation District Permit Review
450–480	County Water and Sewer District Reviews
480–540	State Agency Reviews
390–570	Regional Agency Reviews
280–390	Federal Agency Reviews
330–390	Citizen Participation Program
480–540	HOA Documents, Warranty Registrations

Model Center

480–510	Construction and Sales Sequence Program
430–480	Model Center Design Concept and Approach
430–480	Prepare Model Center Site Plan
430–480	Complete Model Center, Entry Feature Landscaping Plan
480–510	Complete Signage Design
390–480	Complete Interior Design Program
480–540	Obtain Regulatory Permits for Model Center
480–510	Prepare Model Center Budget
480–540	Complete Sales Brochures, Ad Campaign, Billboards

480–540	Complete Scale Model of Project
80–540	Prepare Final Project Proforma and Budget
Construction Process	
480–510	Prepare Land Development Schedule
480–540	Complete Land Development Bid Process
520–540	Complete Land Development Negotiations and Contracts
480–510	Unit Production Schedules
480–540	Unit Construction Contracts
540–630	Model Center Land Development
480–540	Install Temporary Sales Trailer at Site
540–570	Install Sales Displays and Trailer Interiors
540–570	Complete Landscaping and Signage for Sales Trailer
570–570	Preconstruction Sales Opening
590–750	Model Center Construction
720–770	Complete Model Center Landscaping and Hardscape
720–770	Complete Model Center Model Interiors
780–780	Grand Opening
Project Development	
As Needed	Complete Special Land Development Projects
To Completion	Land Development
To Completion	Sales
To Completion	Unit Construction
To Completion	Closings
To Completion	Municipal Acceptance of Infrastrucutre
To Completion	Homeowners Association Transition

Appendix 6-2

MULTIPLE PROJECT MASTER SCHEDULE

		Land Plan		Engineering		Regulatory Process				Land Development				Unit Construction		Comments
# Units	Project	Start	Comp	Prelim Comp	Final Comp	Prelim Submit	Plan Appr	Final Subm	Plans Appr	Contract Comp	Start	Pad	Comp	Start	1st CO	
13	Mark Mills	6/7	7/7	8/17	8/17	8/18	2/9	8/18	2/9	3/9	4/9	7/9	12/12	7/10	12/10	

Appendix 6-3

INDIVIDUAL PROJECT MASTER SCHEDULE

Project:____________________________________	DUE DATES
Project Planning and Subdivision Concept	4/02
File for Wetlands Permits	4/02
Start Preliminary Improvement Plans	4/15
Submit Preliminary Improvement Plans and Plat	6/02
Draft Preliminary HOA for Storm-water Facilities	6/02
Submit Preliminary Plans to State DOT	6/06
Development Review Committee Meeting for Preliminary Plans	6/19
Start Preparation of Final Improvement Plans and Plat	6/20
Design Development Plans for Product	6/27
Planning and Zoning Meeting for Preliminary Application	6/28
County Planning Board	7/19
Complete Final Engineering Plans and Plat	7/21
File for State Permits: Environmental, Traffic, Sewer, Water, etc.	7/25
Complete Final Architectural Working Drawings	8/12
Complete Construction Drawing Revisions	9/15
Forward Drawings to Construction Personnel	9/23
Forward Drawings to Sales	9/23
State Permits Due	9/26
Submit Final HOA Documents for State Review	9/26
Submit Final Plat and Improvement Plans to Planning Board	9/26
Development Review Committee Meeting for Final Application	10/06
Planning and Zoning Board Meeting	10/31
County Planning Board Meeting	11/14
Start Land Development	11/15
Record Final Subdivision Plat	11/23
Submit for Building Permits - Models	11/24
Start Sales	11/24
Start Construction of First Unit	1/24

NOTE: If any of the governmental and submittal dates are missed, the project will be delayed one month.

COMMENTS:____

____

____

Appendix 6-4

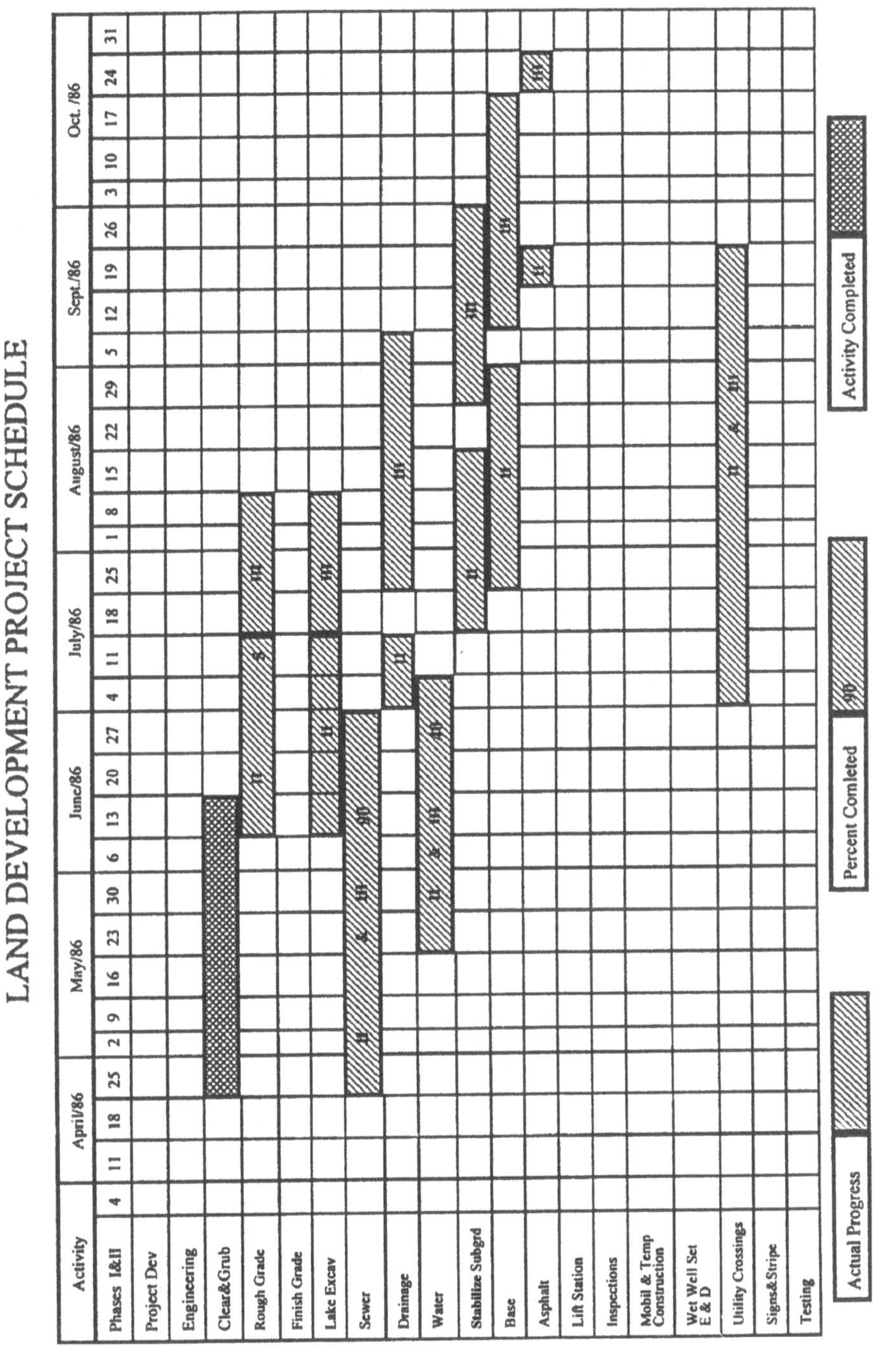
LAND DEVELOPMENT PROJECT SCHEDULE
Activity
April/86
May/86
June/86
July/86
August/86
Sept./86
Oct. /86
Phases I&II
4
11
18
25
2
9
16
23
30
6
13
20
27
4
11
18
25
1
8
15
22
29
5
12
19
26
3
10
17
24
31
Project Dev
Engineering
Clear&Grub
Rough Grade
Finish Grade
Lake Excav
Sewer
Drainage
Water
Stabilize Subgrd
Base
Asphalt
Lift Station
Inspections
Mobil & Temp Construction
Wet Well Set E & D
Utility Crossings
Signs&Stripe
Testing
Actual Progress
Percent Comleted
Activity Completed

Appendix 6-5

HOUSING INVENTORY REPORT

	Starts			Completions			Closings			Inventory Unsold		Inventory Sold		Inventory Total		Net Orders			Order Backlog	
Month	Orig Budg	Actual Forecst	Prior Year	Orig Budg	Act Forcst	Prior Year	Orig Budg	Act Forcst	Prior Year	Comp	Under Const	Comp	Under Const	Comp	Under Const	Orig Budg	Act Forcst	Prior Year	Orig Budg	Act Forcst
Balance	--	--	--	--	--	--	--	--	--	--	--	--	--	49	25					54
Jan	18	3	3	8	5	4	18	14	4	27	7	13	16	40	23	30	10	7	25	50
Feb	29	13	6	8	10	8	20	10	6	30	6	10	20	40	26	41	16	10	25	56
Mar	32	10	8	8	8	3	23	4	6	34	2	10	26	44	28	40	20	20	22	72
Apr	28	27	9	18	7	6	18	5	2	36	1	10	47	46	48	35	18	12	20	85
May	33	16	15	29	7	1	24	10	4	30	2	13	55	43	57	37	20	15	18	95
June	41	14	12	32	12	10	32	15	4	30	2	10	57	40	59	28	16	30	18	96
July	30	8	32	28	16	12	21	16	7	29	2	11	49	40	51	19	11	9	22	91
Aug	30	13	24	33	13	9	21	9	12					44	51	13	14	9	25	96
Sept	35	10	13	41	16	10	26	13	7					47	45	13	16	11	28	99
Oct	10	22	8	30	20	18	23	36	14					31	47	16	26	8	29	89
Nov	10	28	8	30	40	25	33	34	19					37	35	21	28	7	29	83
Dec	10	28	10	35	46	34	41	34	25					49	17	19	26	3	28	75
Total	306	192	148	300	200	140	300	200	110					49	17	312	221	141	28	75

Chapter 7

Basics of Residential Marketing

Successful residential marketing begins early in the development process, at the point where the target market has been identified and the engineers are beginning the project design. In this chapter, engineers—as well as builders and developers—are introduced to residential marketing and learn tips for design and product selection that are applicable for all residential projects.

In a successful subdivision, the housing design not only sets the style for the development but also separates the subdivision from its competitors. The ambiance and amenities of the subdivision should be designed to be memory points for prospective buyers—marketing hot buttons can be incorporated into the design to enhance curb appeal and, ultimately, increase absorption rates. Important marketing factors include location, competition, land plan, lot premiums, housing style, and marketing approach.

The design elements of a subdivision will vary according to the needs and desires of the target market. For example, the lot size and desirable amenities are quite different for the first-time home buyer than they are for the move-down buyer, that is, the retirement market. Moreover, creativity and a perception of the target market's emotional needs are more important to the design than clever engineering or technical innovation. For example, the design of a retirement community should include special attention to the design and land planning for amenities, which usually include pools, clubhouses, tennis courts, special entrance features, security measures, and walking trails, all of which enhance a potential home buyer's sense of safety and comfort.

Sensitive land planning and engineering can bring a project to life, creating a living environment that sells out quickly. Some of the factors that increase market appeal of a new subdivision are considered next: location, competition, land plan, lot premiums, housing selection, marketing approach, and potential for expansion.

Location

It is true that *location, location, location* is the key element of a project that cannot be remedied by money. In site analysis work, location is the starting point for selecting the property. It is also the end point when the project is open for sale. Not only is *location, location, location* important for the marketability of the project, but it is important for the regulatory approval and cost of the project. Connecting to existing infrastructure, such as perimeter roads, water or sewer lines, or drainage facilities, is important to the overall cost of developing the land. However, location is a factor that goes beyond pure dollar value.

Buyers seek subdivisions and housing communities that are well located and close to public services, public offices, shopping centers, and other amenities that add to the quality of life and convenience of everyday living. Engineers providing site selection and analysis services must concentrate on the property location and the site characteristics that enhance the marketability of the site. In terms of analyzing property for purchase deliberation and development purposes, location should be the primary factor in assessing the advantages of introducing a new project to the marketplace.

On this score, infill properties that were previously overlooked may yield great benefits. Overlooked properties have a problem, but engineers can investigate the problem and prepare creative solutions to overcome it. Such properties can be purchased with the knowledge of the constraints and thus under market price. Engineers can add value to the overlooked parcels by designing creatively, so that the marketability of the project is enhanced. It is much more advantageous to satisfy niche markets with overlooked infill properties because the land values should be below market price.

Competition

A good practice for builders is to situate a new project close to the competition. The traffic generated from a competitor's print media, radio advertisements, and direct mail campaigns will attract buyers to its project, which then spills over to adjacent or nearby projects. However, once the buyers visit your site, it is important to differentiate your project from your competitor's. Differences can be drawn by materials, product, color, presentation, or orientation price, but your most important one is a street scene that emanates comfort.

If your project is better located, better priced, and of better value, being close to the competition will result in higher absorption rates for you and ultimately a more profitable project. For this reason, an infill site may be a better selection than a rural setting for your marketplace.

Land Plan

The pricing structure for builders does not vary significantly. Large regional builders may have some benefit in their buying power, but their size does not necessarily make their subdivision design more creative. The pricing of homes or lots directly correlates to the subdivision design and the perceived value of a creative land plan. The sales price of a home does not have a direct relationship with the cost of building it. The market sets the price of the home or lot and hopefully the builder/developer will realize a profit on the land and, with luck, an additional percentage from the house, lot premiums, and extras. Proper land planning for the marketplace and the selection of the hot button amenities for enhancing the site features can increase the perceived value of the project and yet not necessarily add to the overall cost. This reinforces the need to present an imaginative land plan and prepare a conceptual cost estimate before the purchase of the property. The highest and best use of the discretionary funds should be determined at the conceptual land planning phase.

The land planning techniques that result in high lot yields, creative open spaces, and environmentally sensitive land consumption will be more readily accepted by the public, making it easier to market the project. However, new housing ideas may not be readily accepted by the home buying public. Builders want to be leaders but not pioneers in a marketplace. They will not introduce zero lot line or small single-family lots in a marketplace when half-acre lots are the norm—unless, of course, the market research indicates a need for a lower price point and the government promotes smaller lots. Then, builders should seriously consider the idea as long as the market potential and depth are strong. As an engineer, you must demonstrate to your clients that you understand the marketplace and the creative land planning approaches that can be successfully implemented. In most markets, a typical subdivision with open spaces will succeed because that is what the home buying public perceives as a quality subdivision. It will succeed also because most local government standards are not flexible enough to allow innovation in land planning. Without question, the larger the project size, the easier it is to maximize the number of units and be able to afford the ambiance of a creative subdivision design. However, smaller parcels adjacent to or within exclusive communities or even small infill sites may provide opportunities for small to medium size developers. Especially, infill sites provide land planning opportunities to change underutilized land into a contributing tax base. Also, existing infrastructure would benefit a creative, innovative, and cutting-edge project approach.

Lot Premiums

When the housing market is thriving, builders will slow down the absorption rate for the more marketable lots. All buyers want the best location, especially when the best lots have views, woods, and privacy or are located at the end of a cul-de-sac. To slow down the sale of the best lots, the builder can either hold the lots off the market, which is a negative approach, or place premiums on the lots. This slows down the selection of the best lots and allows the typical lots to be sold and constructed first. The philosophy is that the best lots will always sell.

Land plans create lots that can be marketed with lot premiums. These lots will have unique characteristics that are distinctly different from the typical subdivision lots. Analyze each individual lot within the project to ensure the building footprint is consistent with the builder's house line and to provide flexibility to sell any of the homes on each lot, thus permitting the builder to file for a prototype building permit.

Cul-de-sac roads and lots are inviting to home buyers. Cul-de-sacs are always the first lots to be sold even with lot premiums. These lots provide a sense of privacy, security, and safety for the homeowners. In most instances, the elimination of through traffic within a subdivision will be beneficial to the overall presentation of the project.

There seems to be a trend that regulatory planners want to reduce or eliminate cul-de-sacs. Their opinion is that connectivity, which involves connecting streets rather than terminating streets in a cul-de-sac, is more desirable. Their goal is to legislate the interaction between homeowners and prevent isolation of a group of homeowners at the end of a cul-de-sac. This approach seriously miscalculates the marketplace, becoming one more misguided planning concept implemented for cars rather than for people. In fact, all market segments, from the first-time home buyer to the retirement buyer, prefer cul-de-sac lots. The design community should not waiver from the preferences of the marketplace. For many years, developers have been citing the fact that "all of the good land is gone." In this case, *good land* is defined as well located and easy to develop. Based on land forms, cul-de-sacs will never disappear.

Design a land plan for a maximum number of premium lots by using the land characteristics and land constraints to determine the ultimate marketability of the project. Most builders do not include lot premiums in their pro forma, even though they add to the success equation.

Housing Selection

Engineers should know the type and style of houses that are selling in their marketplace. Do not stop at the front walk, either. Walk through your competitors' models to view the presentation inside and outside of the house. The construction techniques used by builders in your marketplace are also important to understand. An engineer can enhance client relationships and design a better-quality subdivision by understanding how the building industry functions and markets its products.

Selection of a foundation is a function of the marketplace but also of the characteristics of the land. An engineer must be the advocate of the land and explain how the land can add value and increase absorption. Select housing and foundations both for the land and the consumer.

Many builders follow the marketplace and offer product selections that are similar in size and style to those of other builders. To differentiate its product, then, the builder changes the product specifications inside the house. In a hot market, this may be the only marketing change necessary. However, in a slow market, interior changes are not enough. Thus, engineers must add perceived value to the subdivision design and land plan all the time. Too often, it is easy to forget the down-market scenario. Time-honored planning and design concepts can add perceived value to any subdivision regardless of the marketplace or market. These concepts include house and site landscaping, gated communities, entry features, design emphasis on people rather than vehicles, lot orientation, privacy, and maintenance considerations.

There are times when preplanning the house type for each lot becomes necessary; for example, foundations should be constructed in the late fall for winter housing production. On the one hand, preplanning denies consumers flexibility in the selection of the home or lot, but on the other hand, the builder knows the costs and can increase efficiency in production. Salespeople balk at preplanning, but engineers must keep the need for preplanning in mind and learn the phasing requirements for each project. The phasing of the number of lots released for sale and the cost of the infrastructure are always important. Design the subdivision so that your builder/developer limits the cash flow, increases marketing flexibility, and can stop the project at well-defined stages. If the builder chooses not to stop at the phase lines and continues the infrastructure improvements to meet the market demand, then the builder's confidence level is high and an increase in sales absorption must be realized to stay in line with the infrastructure improvements. However, the decision of the builder should not affect the way the engineer designs the subdivision infrastructure.

Engineers need a keen understanding of the *actual* cost of developing property and the cost ramifications of the subdivision design features, but they must be willing to go beyond the road right-of-way in their design approach and their understanding of the marketplace.

Marketing Approach

In most markets, a model home is required for buyers to inspect before they purchase. A model home presents confidence that the project is "real." The quality of the products, construction, and presentation provide first and lasting impressions that will make or break a project.

Engineers must understand the purpose of models and model home complexes. When selecting the proper model home location, consider the following factors:

1. Cost to install the infrastructure to serve the models;
2. Timing the model center construction to commence as soon as possible;
3. Orientation of the homes on the lots;
    - For curb appeal: How will consumers view the house when they first drive up to the model center?
    - The view outside from inside of the model: What will consumers see from all the windows?
    - The natural light inside of the house during the entire day.
4. Sales trap within the model center (have home buyers visit the models by coming into and out of the sales area);

5. Arrangement of the model homes in terms of pricing (start with the standard home and walk to the largest model, which offers increased sales and perceived values);
6. Model center design (to enhance excitement in the buying process);
7. Creation of a sense of urgency; and
8. Landscape, hardscape, walkways, lighting, and signage.

The model center and model homes set the tone for the project and the acceptance of the project by the marketplace. The engineer needs to incorporate the model center ideas early in the land planning phase of design. It is also appropriate for the engineer to become involved in the model center beyond the normal grading, drainage, and utility design elements.

Other elements of the marketing approach for a project include sign locations, entrance features, and other elements that give the project a sense of identity and a memory point for the home buyers. Consider different paving materials at entrances, even for public roadways, to provide a sense of arrival to the project. The successful engineering firm in the residential industry thinks like a builder/developer and becomes a leader in the project design team.

Subdivision Potential

During the site analysis phase, investigate the development potential of adjacent properties. It makes sense for a builder to continue building operations in the same location. Subdivision design should accommodate future infrastructure extensions to adjacent properties. Ultimately, engineers have a positive effect on the market's acceptance of a project. Through flexible land planning and creative engineering design, the engineer controls the cost of development, affects the marketability of the lots, and establishes a sense of community within the project. Understanding the residential marketing strategies of the building industry will assist the engineer in better serving the client base and in creating memorable subdivision designs.

Chapter 8

Consultants

Each design professional involved in the process of developing a new subdivision can contribute to a better living environment for future homeowners. Each design criteria, however, must be relevant to housing projects—not commercial, industrial, or highway purposes. Most design professionals, regardless of discipline, are required to take a design course but most learn through experience and experience alone. They use what they have learned from past unsuccessful experiences to shape their design approach for their next housing project. After all, residential land planning is an art and a science combined to create a design style. This design style is perfected over years of practice.

The physical attributes of a site are the basis for the creative design approaches and patterns that make each housing project unique. The proper use of the land will, over time and after many projects, establish the reputation of the developer and design team. Successful housing design professionals are sensitive to site characteristics and how to use the land to benefit the eventual user (Figure 8-1).

Talented design professionals may work out of a garage or basement or they could be members of the largest regional design firm. The difference between large, local, or even regional firms and home-office professionals is not their design abilities but the overhead costs of providing the design service. The design professional best suited for designing a subdivision is the person who understands the factors controlling the housing development process. When selecting a design professional, start with an investigation at the planning and zoning office to determine the designers hired by the competition.

Keep in mind that just because these firms are doing work in the community does not mean that their work is of high quality. Review project plans at the planning office and then compare the approach by visiting the constructed subdivision. Determine whether the project was constructed consistently with the plans. Also, find out whether the design firm is well received at the planning office. Reputable firms with qualified personnel are necessary to obtain permits and approvals

Figure 8-1. Design team. The selection of project team members is based on the design capabilities necessary to meet the street scene development goals of the developer.
Source: David E. Johnson.

in a timely manner. Their ability to understand the political and governmental influences on housing will help shorten the approval process.

Selection Process

Residential subdivision design work requires a unique talent of sensitivity to the way people intend to live.

The following criteria can guide you in the selection of members of the design team:

- Cooperation. Does the consultant have the ability to use ideas generated through project discussions and implement those ideas with enthusiasm?
- Knowledge. Does the consultant have specific knowledge to prepare a design consistent with the intent of the project?
- Workforce. Does the firm have sufficient employees to handle their current workload and manage the proposed project? The project

should receive the same attention as any similar project. Any less attention would indicate a staffing problem.

- Experience. Evaluate the educational background and work experience of the personnel assigned to the project. Visit existing projects designed by the consultant. Is the project appealing? Do the homes complement the land form?
- Commitment. During the discussions with the design team, will the consultant make a commitment to adhere to the project schedule? A personal commitment to a project schedule is important so that the regulatory submissions are made on time.
- Fees. The fee and negotiated contract provisions should reflect a lump sum for specific scopes of work. Does this approach define the work and identify all the costs expected during the design and regulatory process?

The contract agreement with the design professional should be based on a specific time frame for the plans to be completed and submitted to the regulatory agency (Appendix 8-1). Not all design professionals will work on every housing project. For a single-family traditional housing project, the following disciplines should be considered for the design team: civil engineer, environmental consultant, architect, landscape designer, surveyor, and soils engineer. Other professionals to be considered would be an economist to determine the effect of the project on municipal services, schools, fire, and first aid capabilities; a traffic engineer to determine the effect the project would have on existing roadway systems; and a planner to prepare the conceptual site plan. If managed properly, the land planner, who determines the highest and best use of the land, can be the most cost-effective professional retained by a developer.

Design Team Members

The success of a housing project is measured by the ability of the project design team to obtain government approvals and permits in a time frame consistent with company objectives. Ask each design professional to commit to a production schedule for the project. The following outlines the work responsibilities for each design professional and can be used as a guide when discussing work scopes.

Surveyor

At the very beginning of every project, the surveyor provides a boundary survey, a task that includes researching deeds to investigate all easements and encumbrances affecting that property. The survey should detail the existing infrastructure in and around the project site (e.g., roadways, drainage facilities, telephone poles, fire hydrants, and other utilities) that could affect the ultimate layout of the property. In addition to the boundary survey, a complete topography survey is required. The topography is obtained by aerial photography or by an actual field survey. If possible, consider paying for low-flying aerial photography to ensure an accurate topographical survey. The boundary and topography surveys are essential elements of the base map used for land planning.

Environmental Consultant

The environmental consultant's scope of work for site inspection varies from project to project and from one state to another. The US Army Corps of Engineers (USACE) is charged with the federal responsibility for wetland protection; some states have expanded the USACE's role in the protection of wetland areas. The environmental consultant should walk the site, take necessary soil borings, inspect and identify vegetation, analyze the soil, and determine the depth of the groundwater to ascertain the presence of wetland areas. Wetland areas should be flagged in the field and ready for the surveyor to delineate by metes and bounds on the base map. The environmental consultant should know about endangered species and endangered vegetation that are protected by federal, state, and local laws. Other environmental restrictions that should be reviewed and included in this consultant's final report are archeological areas, recreation requirements, type and size of significant trees and stressed vegetation found on the site, floodplains, and so forth. A list of the soils affecting the ability to develop the project should also be incorporated into the final environmental report.

Soils Engineer

The soils engineer manages soil testing rigs and the apparatus needed to dig test pits and undertake a soil boring program for each site under consideration. This investigative work validates characteristics of the site and correlates the data with the soil conservation district soils map. The information provided in the soils report includes the following:

- Definition of the soils on site;
- Capability of the soil to be used for structural fill, foundation work, road construction, and septic systems;
- Analysis of the topsoil;
- Location of unsuitable material that must be supplemented with other types of material; and
- Identification of the groundwater elevation and determination whether the installation of utilities will require dewatering (an added expense).

For residential subdivisions, a grid-patterned soil analysis program is not always necessary. In fact, in many cases, it is a waste of time and effort to test areas located in open spaces or the lots where buildings will not be located, that is, setback areas or buffers. Thus, the soil testing program should concentrate on the center portions of the site where the significant housing and roadway activities will occur. The soils engineer outlines a program for identifying the soil on site and provides recommendations on the benefits and use of the site soils. In rocky or low areas, or areas with unsuitable soils, the soil testing program should be expanded to ensure the site is fully analyzed. The results should be discussed with the land development contractor. Removing soil unknowns from land development contract negotiations enhances the ability to accurately predict costs.

Land Planner

Land planning for residential subdivisions can be mastered by the builder/developer or contracted at minimum cost to a professional land planner. Sometimes a concept plan that is prepared in-house helps company executives understand a property's potential before they decide to retain a professional land planner.

In some states, the land planner is a registered and licensed professional. However, in most locations, the land planner is the architect, landscape architect, or engineer for the project. In the public sector, the skills of a land planner are effectively used to develop community master plans, zoning maps, and plans for capital improvement projects. In residential housing projects, the land planner works to consolidate all the design intentions in a subdivision land plan that adheres to the

Figure 8-2. Land planning. Site characteristics, zoning regulations, and design creativity establish advantageous marketing. Smaller lot sizes allow more open space dedication.
Source: David E. Johnson.

requirements of the zone. The land planner may have to develop several different schemes before the project team selects one that meets the company's goals. Before design plans are submitted for approval, all the design issues should be addressed, and the number of variances or design waivers should be limited to minimize political scrutiny of the project.

The land planner's mission is to maximize the density and adhere to the highest and best use for the property while achieving high project marketability (Figure 8-2). Request that the land planner prepare a loose conceptual lot and street layout, which would subsequently be revised to meet all the technical requirements imposed by local government. If a land planner is retained to prepare site plans, do not restrict the planner's creativity. New, fresh planning ideas can provide a marketing edge. The land planner will also lay out the basic elements of the project site, that is, road alignments, lot configurations, open spaces, walkways, and tree-save areas. The other design professionals then accept the land plan and modify the concept to meet the technical requirements of the municipality.

Engineer

The engineer is the primary designer of subdivisions. Engineers prepare the subdivision design on the basis of all the conceptual drawings and all the residential design criteria. The lot layout and street rights-of-way are hard-lined on the conceptual plan. The engineer assigned to the project must be a registered engineer (and not an engineer-in-training) in the state in which he or she practices. Ask to have an engineer with housing experience be assigned to the project. Review the engineer's work before hiring.

The engineer's role is to design vertical and horizontal alignments of the internal roadways; access points onto existing roadways; water, sewer, and drainage systems for the project; and the grading of each lot. A successful engineering approach (Figure 8-3) should minimize

Figure 8-3. Engineering design. Land development projects require a civil engineering design discipline to reduce site costs and for regulatory compliance.
Source: David E. Johnson.

- Design variances from regulatory criteria,
- Unusual construction costs,
- Complex engineering design to maximize quality control, and
- Awkward lot layouts to maximize land values for marketing success.

Landscape Architect

The landscape architect provides significant insight on the full utilization of the existing attributes of a site. The landscape architect supplies information on tree-save areas and identifies specimen trees that should be saved; doing so will save on clearing costs and establish premium lots. Landscape architects also provide design input regarding irrigation systems, project identification features, street tree planting, housing landscape packages, color coordination, product orientation, and selection of ground cover and grasses. The landscape architect's contribution is to visualize the street scene and ensure the site features are accentuated.

Architect

A competent housing architect can use the land plan and marketing information to provide a design development program that produces sales consistent with the competition's absorption rates. The architect should be consulted on the selection of the standard foundation style of each house, that is, walkout basements, in-ground basements, front-to-back or side-to-side split levels, crawl spaces, and so forth. The engineer uses this information for site grading purposes. The architect uses the site analysis report to prepare housing plans, which use the ground, views, and topography to the benefit of the buyer of the lot or the home. The sales price range of the product line dictates the budget for houseline construction. The architect uses this financial constraint to ensure the designs are marketable and can be constructed within budget.

Other Disciplines

Based on project complexities and the need to supplement full-time staff, builder/developers may consider retaining the services of the following professionals:

- Economist (or the land planner) prepares a fiscal analysis of a project site and its economic effect on schools, emergency services, fire departments, and so on, when these are necessary. The economist should know about municipal services and community school system. This analysis should demonstrate that the new homes will generate more in new taxes than they will cost in terms of community services to serve the homes.
- Real estate agent provides testimony and background on housing prices and verification of the market study.
- Interior designer provides advice on color schemes, architectural layouts, project identification features, marketing schemes, and model decorating.
- Attorneys are key project members during the land purchase and governmental permitting processes. The complexities of real estate law require legal representation for each project. Retaining the services of a local attorney is a prudent business decision to ensure corporate and personal protection under local, state, and federal laws.

The engagement of any consultant should be based on the needs of each particular project. Do not hesitate to hire a consultant who can positively influence the approval, construction, sale, or closing of a project (Figure 8-4). Any housing project must have, at a minimum, a surveyor and an engineer. The other disciplines could be retained as needed before and during the regulatory process. As you would with any consultant, communicate the project objectives, outline your expectations, and seek a commitment to a schedule. All consultants should be sensitive to the social environment of the subdivision, its functionality, livability, and longevity. The consultant is a tool of success in the housing industry—use consultants to attain market acceptability of the subdivision and product.

Professional Service Agreement

The professional service agreement for consultants is provided in Appendix 8-1. This agreement is a base contract agreement between the builder/developer and any design professional. The general provisions of this agreement are common for all consultants and for any project. Once executed by a consultant, specific project contracts can refer back to this base agreement for general provisions, such as

- Ownership of plans;

Figure 8-4. Consultants. Creative design and balance between architecture, landscaping features, and infrastructure elements are cost-effective innovations.
Source: David E. Johnson.

- Compliance with regulations imposed by local, regional, state, and federal governments;
- Indemnification clause;
- Professional liability and errors-and-omissions insurance policies (an insurance certificate must be submitted to the owner and kept on file for reference); and
- Termination clause for both the builder/developer and the design professional.

For specific projects, use the general contract language and scope of work descriptions outlined in Appendixes 8-2 through 8-5. These contracts are for site feasibility and surveying, land planning, engineering, and architecture, respectively. This basic information provides builder/developers with a basis for negotiating contracts with design professionals. The scope of work can be modified to meet the needs and requirements of each specific site and project.

A detailed description of the scope of work for the surveying that is required to prepare a subdivision plat for recordation purposes is outlined in Appendix 8-6. This description of services includes the boundary and topographic surveys. The scopes of work outline the requirements for the surveyor to record the final plat, establish field controls, provide land development construction layout, and provide houseline surveying. Appendix 8-7 is a detailed scope of services for housing construction surveying.

The agreement for architectural services should complement the engineering work and regulatory process. Architectural services include design development work to establish the design program and housing product for the project. Once completed, the preliminary design phases and final construction documents are prepared in accordance with the scope of work.

The construction documents must be signed and sealed by a registered architect. Signing and sealing are requested of the architect when the building department requires it, but builders occasionally ask for this for their own use. The signing and sealing of construction documents establish the liability for the content of the construction documents and elevates the importance of the documents in the field. Some states do not require that one- and two-family residential construction plans be prepared by a licensed architect. Some states, however, require certified designers to prepare the plans for building permit purposes.

Consultant Responsibility

If a consultant is retained for design services, the builder/developer should use the contract agreement to ensure that the construction documents is consistent with all governmental regulations and is of sufficient detail to obtain building permits. By hiring design consultants knowledgeable in residential housing construction, a builder/developer hopes to ensure that the plans will be sensitive to the ultimate user and that regulatory permits will be obtained in a highly efficient manner.

When the project is under construction, the plans should not be open for interpretation. There will always be construction variations, construction mistakes, or design changes that should be approved by the design professional. All deviations from the plans should be approved by the design professional to ensure his or her professional liability does not become null and void. The prepared contract documents become the basis not only for all bid considerations and cost estimates but for the future selection of design professionals. The quality of the construction documents is the best guideline for assessing the ability of the land development contractor and the potential success of the project.

Because design professionals should sign, seal, and stand behind their design work, occasionally consultants find it convenient to use government regulations to justify their design decisions. However, it is *always* incumbent on design professionals to use those regulations in a context that best benefits the project as a whole. If the project schedule can be managed properly and the closing schedule will not be jeopardized, designer professionals should exercise their freedom to select standards that conflict with existing regulations but that enhance the project's approach, aesthetic appearance, and function.

Design professionals must balance the special interests of the project, including the government, client, demands of producing a subdivision meeting the market, and their own professional ethics and design philosophies for residential projects. The successful designer is able to blend all of these considerations into an approach that can be efficiently approved, effectively marketed, and satisfactorily constructed.

Design professionals must be advocates of the home buying public, who is not present to support new housing projects at regulatory hearings. Using market information and a knowledge of government regulations as guidelines, the professional housing designer prepares a subdivision that ultimately will meet the needs of new home buyers.

Consultants are in business to earn a living, and the greater the workload or number of projects, the more the potential for financial success is possible. Housing projects require attention to detail that can be lost when a consultant strives to finish one project too quickly in order to start another.

Builders and developers must assess the capabilities of the consulting design professionals to ensure that their project is receiving proper staff attention. Builder/developers must hold design professionals to a very high standard—once the residential housing project has been completed, the reputation of the builder/developer will be evaluated by government and the general public, not the consultant or the land development contractor. The residential housing industry has always been based on business and personal relationships. The synergy between the developer and design engineer are tantamount to a successful project. Building long-term relationships with design team members and the building community benefits the community in which they work. The end user of a project is the direct beneficiary of a strong business relationship among all entities.

Appendix 8-1

PROFESSIONAL SERVICE AGREEMENT

THIS AGREEMENT made this___day of__________by and between______, having an address of____
________________________________, hereinafter called "______" and______________
hereinafter called "CONTRACTOR".

For and in consideration of the mutual covenants herein contained the parties hereto do hereby agree as follows:

The Contractor agrees, as an independent contractor, to furnish such labor and/or equipment and/or materials and/or to perform such services as may be specified in any Addendum Schedule executed by the parties and issued in connection with this Agreement.

All designs, drawings, specifications, technical data, and other instruments produced by Contractor in the performance of this Agreement shall be the sole property of and is vested with all rights therein of whatever kind and however created, whether by common law, statutory law, or by equity. Contractor agrees that shall have access at all reasonable time to inspect and make copies of all notes, designs, drawings, specifications and other technical data pertaining to the work to be performed under this Agreement.

Contractor shall comply with existing ordinances and rules of municipality where work is done, and shall comply with all County, State and Federal laws pertaining to such work. Contractor shall also comply with all State Safety Ordinances, Federal OSHA Standards and Company Programs.

The Contractor agrees to and does hereby indemnify and save harmless from and against any and all loss, damage, claims, actions, liability and expense in contract or in tort, in connection with loss of life, bodily injury and/or property damage occurring on or about or arising out of those areas of the Project occupied by him or under his control, or wherever arising if occasioned wholly or in part by any act or omission of, or any defect in workmanship, services or material furnished by the Contractor, his subcontractors, sub-subcontractors, or any subcontractors or sub-subcontractors of any of them, or by his or their agents, servants, employees or materialmen, whether the same arise during the progress of the work or at any time subsequent to completion of the work.

"Contractor", shall provide and maintain in full force in connection with performance of insurance operation Worker's Compensation Insurance, comprehensive General Liability (including but not limited to, contractural products/completed operations including broad form property damage and personal injury coverage), with limits of $1,000,000 Combined Single Limit (CSL) for both bodily injury and property damage. An umbrella policy with a minimum limit of $2,000,000 must be excess above the underlying $1,000,000 CSL. In addition, automobile liability must be maintained with a $1,000,000 CSL Limit and be scheduled in the umbrella policy. All insurance shall be in force until expiratoin of the contract agreement. Said policies shall contain a provision that the same will not be cancelled or materially changed without thirty (30) days prior written notice to. Contractor and/or contractor's sub-contractor shall deliver to certificates of the workmen's comprehensive general liability and the legal liability property damage policies showing due issuance of such insurance. Contractor agrees that business operations shall not begin unless and until such certificates covering same are delivered.

An engineer, architect, surveyor or any other outside professional service contracted by must provide professional errors and omissions insurance with a minimum of $1,000,000 limit and must maintain this coverage for two (2) years after the completion of said work. All other conditions set forth must also be evidenced in addition to professional liability.

Insurance Company must be at minimum an AM Best "A+" rated carrier and licensed to do business in the State of .

Contractor is engaged by only for the purposes and to the extent set forth in this Agreement, and his relation to, during the term of this Agreement, shall not be considered under the provisions of this Agreement as having an employee status or being entitled to participate in any benefits for regular employees.

Either party hereto may terminate this Agreement for any reason upon THIRTY (30) DAYS prior written notice given to the other party by certified or registered mail and sent to the address of said other party indicated hereinabove.

Any waiver, alteration, or modification of any of the provisions of this Agreement shall not be valid unless in writing and signed by the parties.

IN WITNESS WHEREOF, the undersigned parties have hereby caused this Agreement to be duly executed on the day and year first above written.

COMPANY:________________TITLE:______________________DATE:_________
CONTRACTOR:_____________TITLE:______________________DATE:________

Appendix 8-2

SITE FEASIBILITY AND SURVEY CONTRACT ADDENDUM

XYZ CORPORATION ADDENDUM #____________
STREET ADDRESS DATE:__________________
CITY, STATE, ZIP PAGE:______ OF _________
FORMING PART OF CONTRACT # __________________ DATED:__________
BETWEEN XYZ AND____
FOR WORK AT____

CATEGORY OF WORK: SITE FEASIBILITY AND SURVEY

NOTE:THE STATED PRICE BELOW INCLUDES ALL FEDERAL, STATE AND LOCAL TAXES

<u>DESCRIPTION AND SCOPE OF WORK</u>:
The consultant named above shall perform the following tasks for ____________________ ____________________ project at <u>City, State</u>. This scope of services limits itself to a site feasibility study, survey and preparation of an Exhibit "B" for the Department of Community Affairs, including a site analysis with consideration of XYZ's projected type of housing product.

XYZ Corporation is considering a ____________________ housing product. The size of the site is approximately+/- ac.

With all of the explained above, the consultant shall proceed to:
<u>Task I - - Survey</u>
A. The Consultant shall perform a boundary and topographic survey including, but not being limited to, the following:
   1. Existing elevations on site
   2. Location and elevation of any existing swales
   3. Location and elevations of any existing ponds
   4. Water surface elevations of the adjacent waterway
   5. Elevation of Quesada Avenue, at center line, edge of pavement and R/W line
   6. Location and elevations of any drainage structures
   7. Location of all existing utilities

The survey shall extend 50' beyond the site boundaries and shall be included on a sketch at 1" = 50' scale. (1 ft. = 30 centimeters; 1 acre = 0.4 hectares)

CONTRACTOR:________________ XYZ CORPORATION:____________________
BY:______________________________ BY: _______________________________
TITLE:________________ DATE:_______ TITLE:__________________DATE:___

PREPARED BY:_____________________________________ DATE:__________
REVIEWED BY:_____________________________________ DATE:__________

CONTRACT ADDENDUM

XYZ CORPORATION ADDENDUM #____________
STREET ADDRESS DATE: ___________________
CITY, STATE ZIP PAGE:_______OF_________
FORMING PART OF CONTRACT # ________ DATED: __________________
BETWEEN XYZ AND ____
FOR WORK AT____

CATEGORY OF WORK: SITE FEASIBILITY AND SURVEY

Task II - - Site Analysis

A. Some of the following data shall be collected from XYZ, or any public office in cases where these would not be available at XYZ. Others will be reflected on the survey performed in Task I :

1. Comprehensive Plan*
2. XYZ Land Plan*
3. Vicinity Map*
4. Zoning Map*
5. Topography*
6. Aerial*
7. Utility Plans*
8. Road and Drainage Plans*
9. Transportation Information
10. Subdivision Regulations*
11. Flood Plain Information (FIA Maps)
12. Soils

*Available information at XYZ.

B. A meeting between the consultant and XYZ's Community Director shall be arranged by XYZ. A determination of the availability of services, agency/community attitudes, regulations, requirements and an onsite physical investigation of the property shall be a part of this meeting.

After A and B have been completed, a thorough analysis shall be effected by the consultant and implemented into TASK III.

CONTRACTOR:______________________ XYZ CORPORATION________________
BY:__________________________________BY:______________________________
TITLE:________________ DATE:______TITLE:_________________ DATE:____

PREPARED BY:____ DATE:_______
REVIEWED BY:____ DATE:________

CONTRACT ADDENDUM

XYZ CORPORATION ADDENDUM #____________
STREET ADDRESS DATE: ____________________
CITY, STATE ZIP PAGE:________OF__________
FORMING PART OF CONTRACT # ___________ DATED: __________________
BETWEEN XYZ AND ____
FOR WORK AT____

CATEGORY OF WORK: SITE FEASIBILITY AND SURVEY

TASK III -- FEASIBILITY REPORT

Use site analysis information and survey data to evaluate the site for development potential. Prepare written report which shall include, but not be limited to, the following information and supplemented by key exhibits, maps or plans.

1. Location of the property
2. Legal description of the property
3. Size of property in acres
4. Brief background history of the land
5. Development status (according to D.S.P./XYZ agreement of ______________)
6. Projected number and density, based on desired product by XYZ. Also, discuss procedure of subdivision regulations and site plan review process, including time frames.
7. Define what portions, if any, of the land may be excluded from computing the net buildable acreage.
8. Length of frontage; depth of parcel
9. Zoning requirements; setbacks and other open space required:
   a) Building restrictions
   b) Architectural restrictions
   c) Local restrictions
10. Existing and anticipated easements or rights of ways affecting the property. Provide source of information.
11. Evaluation of available topography and fill requirements
12. Evaluation of soils. Soil tests will be provided by XYZ.

CONTRACTOR:__________________XYZ CORPORATION_____________________
BY:__________________________________BY:______________________________
TITLE:__________________ DATE:______ TITLE:__________________DATE:_____

PREPARED BY:____ DATE:________
REVIEWED BY:____ DATE:_________

CONTRACT ADDENDUM

XYZ CORPORATION
STREET ADDRESS
CITY, STATE ZIP
FORMING PART OF CONTRACT # ___________ DATED: ______________
BETWEEN XYZ AND ______________________________
FOR WORK AT______________________________

ADDENDUM #___________
DATE: ______________
PAGE:_______OF________

CATEGORY OF WORK: SITE FEASIBILITY AND SURVEY

TASK III -- FEASIBILITY REPORT (Cont'd)

13. Description of existing vegetation
14. Impact of Floodplains: Construction problems, filling, etc.
15. Water Management (storm drainage) requirements, and identification of agencies with jurisdiction:
    a) On-site
    b) Off-site
16. Water & Sewer:
    a) Check availability and utility plant capacity
    b) Discuss if any special improvements are needed
    c) Consideration of a sewer package treatment plant and its approximate cost
    d) Consideration of a water package plant and its approximate cost
17. Other Utilities: Check availability of gas, telephone, power and cable TV, and identify supplying companies
18. Accessibility
    a) Condition, ownership and improvement requirements of frontage road
    b) Existing and anticipated traffic flow problem
    c) Acceptability of entrances to access road
    d) Future roads
        1) Right of way needed
        2) New road planned
        3) Connections to existing or future adjacent developments
19. Nature of surrounding land and type of surrounding zoning
20. Anticipated off-site improvements -- estimated costs, if any

CONTRACTOR:____________________ XYZ CORPORATION_______________
BY:______________________________ BY: ____________________________
TITLE:_______________ DATE:_______TITLE:_______________DATE:____

PREPARED BY:______________________________________ DATE:________
REVIEWED BY:______________________________________ DATE:________

CONTRACT ADDENDUM

XYZ CORPORATION ADDENDUM #__________
STREET ADDRESS DATE: __________
CITY, STATE ZIP PAGE:______OF______
FORMING PART OF CONTRACT # __________ DATED: __________
BETWEEN XYZ AND ____________________
FOR WORK AT____________________

CATEGORY OF WORK: SITE FEASIBILITY AND SURVEY

TASK III -- FEASIBILITY REPORT (Cont'd)

21. Environmental concerns and/or impact
22. Attitude of citizens, regulating officials, elected officials and others
23. Preliminary ideas on ways to overcome identified restrictions
24. Outstanding reasons for developing the site
25. Outstanding reasons for not developing the site at this time
26. Recommendation

At a scale adequate for presentation, the written report shall be accompanied by blown up key exhibits, maps or plans. Boards 24" x 36" minimum size are recommended. 1 inch = 25 millimeters

TASK IV -- PRESENTATION

Provide XYZ with 5 copies of the Feasibility Report (TASK III) and 5 copies of D.S.P. Exhibit "B" (TASK IV). Note: For the preparation of TASK IV, the consultant shall use the same exhibits and other information included in TASK III. However, information in the Feasibility Report (TASK III) will include more than required for TASK IV.

FEES

For professional services to be rendered for the above and based on estimated hours to be spent by principal and staff, the consultant shall charge a maximum up-set fee of:

1) TASK I, SURVEY__________ $__________
2) TASKS II., III. and IV __________ $__________

TOTAL $__________

CONTRACTOR:__________ XYZ CORPORATION__________
BY:__________ BY:__________
TITLE:__________ DATE:_____ TITLE:__________DATE:___

PREPARED BY:____________________ DATE:_____
REVIEWED BY:____________________ DATE:_____

CONTRACT ADDENDUM

XYZ CORPORATION ADDENDUM #__________
STREET ADDRESS DATE: __________
CITY, STATE ZIP PAGE:______OF______
FORMING PART OF CONTRACT # __________ DATED: __________
BETWEEN XYZ AND __________
FOR WORK AT__________

CATEGORY OF WORK: SITE FEASIBILITY AND SURVEY

GENERAL:

A. The consultant shall not subcontract out any work outlined in this agreement without prior approval by XYZ.

B. The consultant will assist XYZ through attendance at meetings with regulatory agencies in preparation of the site analysis and feasibility report, if required.

C. The scope of services shall be completed within 15 calendar days upon notice to proceed

D. XYZ reserves the right to stop the work at any moment and remunerate according to the completion of the work up to that moment.

E. Compensation -- XYZ agrees to compensate the consultant for the tasks outlined in the scope of services included herein. This shall be paid by XYZ upon approval of the report, survey and exhibits by XYZ.

F. Consultant to indicate method of payment.

CONTRACTOR:__________ XYZ CORPORATION__________
BY:__________ BY:__________
TITLE:__________ DATE:______ TITLE:__________ DATE:___

PREPARED BY:__________DATE:______
REVIEWED BY:__________ DATE:______

CONTRACT ADDENDUM

XYZ CORPORATION ADDENDUM #____________
STREET ADDRESS DATE: ____________________
CITY, STATE ZIP PAGE:_______OF_________
FORMING PART OF CONTRACT # ___________ DATED: ___________________
BETWEEN XYZ AND ____
FOR WORK AT____

CATEGORY OF WORK: SITE FEASIBILITY AND SURVEY

GENERAL: (Cont'd)

G. Service authorized by XYZ other than those specifically outlined in the scope of services shall be considered additional services. Additional compensation shall be mutually agreed upon by XYZ and the consultant and shall be invoiced at an hourly rate of $________ for principal and $________________ for staff.

H. XYZ shall reimburse the consultant for all out-of-pocket expenses directly chargeable to the project, at actual costs incurred. Such charges shall be itemized and included in the monthly invoices for time charges. Typical reimbursable expenses include travel, lodging, meals, printing costs, long distance telephone calls and other out-of-pocket expenses.

NO GUARANTEE OF THE NUMBER OF UNITS TO BE RELEASED TO THE SUBCONTRACTOR IS MADE OR IMPLIED. AUTHORIZATION TO PROCEED WITH WORK UNDER THIS CONTRACT ADDENDUM WILL BE FORMALLY ISSUED BY THE DIRECTOR OF OPERATIONS IN THE FORM OF A WRITTEN RELEASE OF UNITS FOR CONSTRUCTION. PROFESSIONAL SERVICES WOULD BE RELEASED BY THE DIRECTOR OF PROJECT DEVELOPMENT.

CONTRACTOR:______________________ XYZ CORPORATION:______________
BY:________________________________BY:_____________________________
TITLE:________________ DATE:______ TITLE:_________________DATE:___

PREPARED BY:______________________________________ DATE:________
REVIEWED BY:______________________________________ DATE:________

Appendix 8-3

SITE PLANNING CONTRACT

CONTRACT ADDENDUM

XYZ CORPORATION ADDENDUM #_______________
STREET ADDRESS DATE:_______________________
CITY, STATE, ZIP PAGE:_______ OF ___________
FORMING PART OF CONTRACT # ____________________ DATED:___________
BETWEEN XYZ AND____
FOR WORK AT____

CATEGORY OF WORK: SITE PLANNING CONTRACT

NOTE: THE STATED PRICE BELOW INCLUDES ALL FEDERAL, STATE AND LOCAL TAXES.

DESCRIPTION & SCOPE OF WORK:

The consultant named above shall perform the following tasks for Tracts #, Project Name at City, State. This scope of services limits itself to site planning, with consideration of XYZ's projected type of housing product, and other physical and regulatory constraints.

XYZ is considering ______________________________ housing product. The size of the site is approximately ±____ acres. Maximum allowed density is desired.

All of the above shall be in accordance with all of the requirements of the City of______, in respect to its subdivision and zoning regulations and site plan approval process.

The consultant shall ensure that all of the site planning to be performed shall be consistent, in concept, with the design criteria of the stormwater management agency(s). It shall be the consultant's responsibility to check all the pertinent regulation requirements and ensure that all have been considered and reflected on the conceptual site plan.

CONTRACTOR:__________________ XYZ CORPORATION____________________
BY:_______________________________ BY:_______________________________________
TITLE: _____________DATE:________ TITLE:________________DATE:_________

CONTRACT ADDENDUM

XYZ CORPORATION ADDENDUM #_______________
STREET ADDRESS DATE:_______________
CITY, STATE, ZIP PAGE:______ OF ________
FORMING PART OF CONTRACT # _______________ DATED:________
BETWEEN XYZ AND______________________________
FOR WORK AT______________________________

CATEGORY OF WORK: SITE PLANNING CONTRACT (Cont'd)

With all of the explained above, the consultant shall proceed to:

I. SITE PLANNING

Prepare three (3) planning concepts for the site. These schemes shall be compatible with the overall parameters set by the Department of Project Development of XYZ in respect to the proposed housing product and shall graphically define the land use, including, but not being limited to:

A. Building locations
B. Open space
C. Orientation
D. Amenities
E. Conceptual storm-water management
F. Access points
G. Topographic constraints (If any)
H. Flood prone areas (If any)
I. Other physical problem areas (If any)
J. Easements (If any)
K. Circulation network (Automobile & pedestrian)

NOTES: 1. Each scheme shall be subdivided into four (4) phases, which shall be interconnected with each other, in order to adequately provide for sequential development.
2. Each of the three (3) schemes shall be attractively rendered and colored.

CONTRACTOR:_______________ XYZ CORPORATION_______________
BY:_______________ BY: _______________
TITLE:________ DATE:______ TITLE:________ DATE:______

PREPARED BY:________________________ DATE:______
REVIEWED BY:________________________ DATE:______

CONTRACT ADDENDUM

XYZ CORPORATION ADDENDUM #__________
STREET ADDRESS DATE:__________
CITY, STATE, ZIP PAGE:_____ OF _____
FORMING PART OF CONTRACT # __________ DATED:________
BETWEEN XYZ AND__________
FOR WORK AT__________

CATEGORY OF WORK: SITE PLANNING CONTRACT

DESCRIPTION & SCOPE OF WORK (Cont'd)

3. Each of the three (3) schemes shall state the following information:
   a) Name of project
   b) Location
   c) Existing zoning
   d) Acreage (total and per phase)
   e) Density allowed
   f) Density provided
   g) Number of units (total and per phase)
   h) Area and percentage covered by impervious surfaces (total and specific)
   i) Maximum lot coverage allowed
   j) Maximum lot coverage provided
   k) Open space required
   l) Open space provided
   m) Setbacks required

4. Originals shall be given to the Department of Project Development. The consultant shall keep a set of prints for future reference.

FEES:
For professional services to be rendered for the above and based on estimated hours to be spent by principal and staff, the consultant shall charge a maximum up-set fee of:

$__________

CONTRACTOR:__________ XYZ CORPORATION__________
BY:__________ BY: __________
TITLE:________ DATE:_____ TITLE:________ DATE:_____

PREPARED BY:__________ DATE:_____
REVIEWED BY:__________ DATE:_____

CONTRACT ADDENDUM

XYZ CORPORATION ADDENDUM #____________
STREET ADDRESS DATE:____________
CITY, STATE, ZIP PAGE:______ OF ______
FORMING PART OF CONTRACT # ____________ DATED:______
BETWEEN XYZ AND________________________
FOR WORK AT________________________

CATEGORY OF WORK: SITE PLANNING

DESCRIPTION AND SCOPE OF WORK: (Cont'd)

GENERAL:

A. There shall not be any repeat fees involved in this contract.

B. XYZ reserves the right to reuse the construction documents in any other site or community without having to give any additional compensation to the consultant.

C. XYZ reserves the right of ownership of the construction documents, as defined in Paragraph #2 of the Professional Service Agreement signed between the consultant and XYZ.

D. The consultant shall not subcontract out any work outlined in this agreement without prior approval by XYZ.

E. The consultant will assist XYZ, through attendance at meetings with regulatory agencies in preparation of the construction documents, if required.

F. XYZ reserves the right to stop the work at any moment and remunerate according to the completion of the work up to that moment.

G. Compensation - XYZ agrees to compensate the consultant for the work outlined in the scope of services included herein. This shall be paid by XYZ upon approval of the site plans by XYZ.

CONTRACTOR:____________ XYZ CORPORATION____________
BY:____________ BY: ____________
TITLE:______ DATE:______ TITLE:______ DATE:______

PREPARED BY:____________ DATE:______
REVIEWED BY:____________ DATE:______

CONTRACT ADDENDUM

XYZ CORPORATION ADDENDUM #______________
STREET ADDRESS DATE:____________________
CITY, STATE, ZIP PAGE:_______ OF ___________
FORMING PART OF CONTRACT # ____________________ DATED:__________
BETWEEN XYZ AND____
FOR WORK AT____

CATEGORY OF WORK: SITE PLANNING

DESCRIPTION AND SCOPE OF WORK: (Cont'd)

GENERAL:

H. Consultant to indicate method of payment.

I. Service authorized by XYZ other than those specifically outlined in the scope of services shall be considered additional services. Additional compensation shall be mutually agreed upon by XYZ and the consultant and shall be invoiced at an hourly rate of $_see below_ for principal and $__see below__ for staff.

J. XYZ shall reimburse the consultant for all out-of-pocket expenses directly chargeable to the project, at actual costs incurred. Such charges shall be itemized and included in the monthly invoices for time charges. Typical reimbursable expenses include travel, lodging, meals, printing costs, long distance telephone calls and other outstanding out-of-pocket expenses.

Principal	$______ Per Hour
Sr. Landscape Architect	$______ Per Hour
Landscape Architect	$______ Per Hour
Draftsman	$______ Per Hour

NO GUARANTEE OF THE NUMBER OF UNITS TO BE RELEASED TO THE SUBCONTRACTOR IS MADE OR IMPLIED. AUTHORIZATION TO PROCEED WITH WORK UNDER THIS CONTRACT ADDENDUM WILL BE FORMALLY ISSUED BY THE DIRECTOR OF OPERATIONS IN THE FORM OF A WRITTEN RELEASE OF UNITS FOR CONSTRUCTION. PROFESSIONAL SERVICES WOULD BE RELEASED BY THE MANAGER OF PROJECT DEVELOPMENT.

CONTRACTOR:____________________ XYZ CORPORATION________________
BY:___________________________ BY:____________________________
TITLE:____________ DATE:_______ TITLE:________________ DATE:_______

PREPARED BY:_________________________________ DATE:_______
REVIEWED BY:_________________________________ DATE:_______

Appendix 8-4

ENGINEERING SERVICES

CONTRACT ADDENDUM

XYZ CORPORATION ADDENDUM #_______________
STREET ADDRESS DATE:_____________________
CITY, STATE, ZIP PAGE:______ OF __________
FORMING PART OF CONTRACT # ____________________ DATED:__________
BETWEEN XYZ AND____
FOR WORK AT____

CATEGORY OF WORK: ENGINEERING SERVICES INCLUDING PLAT, PAVING, DRAINAGE, WATER & SEWER

NOTE: THE STATED PRICE BELOW INCLUDES ALL FEDERAL, STATE AND LOCAL TAXES

DESCRIPTION & SCOPE OF WORK:

The consultant named above shall perform the following tasks in accordance with the project. This scope of services limits itself to site engineering including plat, paving and drainage and utilities drawings and calculations, appropriately signed and sealed, necessary for approximately ±48 acres of land proposed for residential use.

The above acreage will have the following housing product:

____

____

All of the above shall be in accordance with all of the requirements of ______County, in respect to its subdivision and zoning regulations and site plan approval process. And also in accordance with all the other agencies with jurisdiction in the project in respect to storm-water management and utilities. It shall be the consultant's responsibility to check all the pertinent regulation requirements and ensure that all have been met and that each submission is complete, with all required items included, so as to obtain all permits.

CONTRACTOR:__________________ XYZ CORPORATION__________________
BY:_____________________________BY: _____________________________
TITLE:____________ DATE:_____ TITLE:__________________DATE:____

PREPARED BY:______________________________________ DATE:_______
REVIEWED BY:______________________________________ DATE:_______

CONTRACT ADDENDUM

XYZ CORPORATION ADDENDUM #________
STREET ADDRESS DATE:________
CITY, STATE, ZIP PAGE:____ OF ____
FORMING PART OF CONTRACT # ________ DATED:________
BETWEEN XYZ AND________
FOR WORK AT________

CATEGORY OF WORK: ENGINEERING SERVICES (Cont'd)

Keeping within the parameters fixed on the schematic plan approved by XYZ concerning the site plan, the consultant shall render the following services:

Task I - Preliminary Design Development

1. The consultant shall prepare the preliminary plat and preliminary site work (including paving, drainage, and utility plans) for the appropriate governmental submissions in accordance with the project schedule.

2. The consultant will attend required public meetings as well as meeting with XYZ, as required, to formalize the design.

3. The consultant shall review the project with the pertinent Federal, State, Regional, and Local agencies and prepare the appropriate applications when required. The consultant will inform XYZ of the requirements imposed by the agencies. Decisions based upon the requirements will be the responsibility of XYZ. The consultant will act in an advisory role during the decision process in evaluating all governmental recommendations based upon their reviews.

4. The consultant shall finalize the preliminary drawings in accordance with review comments and suggestions so as to obtain governmental approvals.

5. The consultant shall design the water and sewer lines within the site and from the site to the existing water and sewer lines. The design shall be in accordance with the local utility standards.

CONTRACTOR:________ XYZ CORPORATION________
BY:________ BY:________
TITLE:________ DATE:____ TITLE:________ DATE:____

PREPARED BY:________ DATE:____
REVIEWED BY:________ DATE:____

CONTRACT ADDENDUM

XYZ CORPORATION ADDENDUM #____________
STREET ADDRESS DATE:____________
CITY, STATE, ZIP PAGE:______ OF ______
FORMING PART OF CONTRACT # ____________ DATED: ____________
BETWEEN XYZ AND____________
FOR WORK AT____________

CATEGORY OF WORK: ENGINEERING SERVICES (Cont'd)

6. Proper coordination with XYZ Engineering and Land Development Departments shall be maintained by the consultant throughout the project for the incorporation of accepted XYZ construction standards.

7. An estimate of probable cost shall be submitted to XYZ at the completion of preliminary design development. This shall meet the construction budget approved by XYZ. In cases where the estimate is over the budget, the consultant shall advise XYZ how to reduce the cost.

Task II - Final Construction Documents

1. The consultant shall prepare final engineering site work (including final paving, drainage, and utility plans), and final plat plans for the appropriate government submissions and make application for necessary approvals in accordance with the project schedule. The engineering design shall be in accordance with XYZ land development standards.

2. The consultant will work with XYZ in preparing for governmental submissions, providing the appropriate documents, applications and attending meetings as required.

3. The consultant shall finalize the construction drawings in accordance with review comments and suggestions so as to obtain governmental approvals.

4. The consultant shall make applications for obtaining the necessary Federal, State, Regional and/or Local permits required. Drawings shall be revised by the consultant, if required, because of non-compliance by any of the agencies with jurisdiction on the project at no extra charge to XYZ.

CONTRACTOR:____________ XYZ CORPORATION____________
BY:____________ BY: ____________
TITLE:______ DATE:______ TITLE:______ DATE:______

PREPARED BY:____________ DATE:______
REVIEWED BY:____________ DATE:______

CONTRACT ADDENDUM

XYZ CORPORATION ADDENDUM #______________
STREET ADDRESS DATE:____________________
CITY, STATE, ZIP PAGE:________ OF __________
FORMING PART OF CONTRACT # ________________ DATED:__________
BETWEEN XYZ AND____
FOR WORK AT____

CATEGORY OF WORK: ENGINEERING SERVICES (Cont'd)

Task II - Final Construction Documents (Cont'd)

5. The final plat shall be prepared to be suitable for recording with bearings and dimensions of all lines of roadways, lots, easements, recreation areas and the legal dedications, according to the rules and regulations of the State of _________.

6. Reviews of progress prints shall be done so between the consultant and XYZ Project Development Department, for proper coordination and compliance of the project with XYZ's goals and objectives.

7. The consultant shall prepare a final cost estimate at the completion of final construction documents, which shall meet the approved construction budget.

8. Specifications shall be included on the drawings.

CONTRACTOR:________________ XYZ CORPORATION________________
BY:________________________ BY:____________________________
TITLE:__________ DATE:_______ TITLE:________________ DATE:_______

PREPARED BY:____________________________________ DATE:_______
REVIEWED BY:____________________________________ DATE:_______

CONTRACT ADDENDUM

XYZ CORPORATION ADDENDUM # ______________
STREET ADDRESS DATE:______________________
CITY, STATE, ZIP PAGE:_______ OF ____________
FORMING PART OF CONTRACT # ____________________ DATED:___________
BETWEEN XYZ AND____
FOR WORK AT____

CATEGORY OF WORK: ENGINEERING SERVICES (Cont'd)

Task III
The consultant shall provide on-site construction observation services on as as-requested basis.

Fee schedule

I. Preliminary Design Development

1. Preliminary Plat	$______________	(Including survey for plat recording)	
2. Civil (Paving, and Drainage Utilities)	$______________		
3. Public Hearings,	$______________		
4. Governmental Approvals & Other Conferences			
5. Estimating	$______________		
Total			$__________

II. Final Construction Documents

1. Final Plat	$______________	
2. Civil (Paving, Drainage, Utilities)	$______________	
3. Public Hearings, Governmental Approvals & Other Conferences	$______________	
4. Estimating	$______________	
5. Specifications	$______________	
Total		$__________

CONTRACTOR:__________________ XYZ CORPORATION__________________
BY:____________________________ BY: ____________________________
TITLE:_____________ DATE:_______ TITLE:___________________DATE:________

PREPARED BY:____ DATE:________
REVIEWED BY:____DATE:________

CONTRACT ADDENDUM

XYZ CORPORATION ADDENDUM #________________
STREET ADDRESS DATE:________________________
CITY, STATE, ZIP PAGE:_______ OF ___________
FORMING PART OF CONTRACT # ____________________ DATED:__________
BETWEEN XYZ AND____
FOR WORK AT____

CATEGORY OF WORK: ENGINEERING SERVICES (Cont'd)

III. Construction Observation

Site Visits $________________

These shall reflect the fees per visit, and shall be as stated in Task III "on an as-requested basis" (reimbursables shall be paid extra).

GENERAL:

A. There shall not be any repeat fees involved in this contract.

B. XYZ reserves the right to reuse the construction documents in any other site or community without having to give any additional compensation to the consultant.

C. XYZ reserves the right of ownership of the construction documents, as defined in Paragraph #2 of the Professional Service Agreement signed between the consultant and XYZ.

D. The consultant shall not subcontract out any work outlined in this agreement without prior approval by XYZ.

E. The consultant will assist XYZ, through attendance at meeting with regulatory agencies in preparation of the construction documents, if required.

F. XYZ reserves the right to stop the work at any moment and remunerate according to the completion of the work up to that moment.

G. Compensation - XYZ agrees to compensate the consultant for the tasks outlined in the scope of services included herein. This shall be paid by XYZ upon approval of the construction documents.

CONTRACTOR:____________________ XYZ CORPORATION________________
BY:_____________________________ BY: ___________________________
TITLE:____________ DATE:_______ TITLE:________________DATE:________

PREPARED BY:____ DATE:________
REVIEWED BY:____ DATE:________

CONTRACT ADDENDUM

XYZ CORPORATION — ADDENDUM #______________
STREET ADDRESS — DATE:____________________
CITY, STATE, ZIP — PAGE:______ OF __________
FORMING PART OF CONTRACT # ____________________ DATED:__________
BETWEEN XYZ AND____
FOR WORK AT____

CATEGORY OF WORK: ENGINEERING SERVICES

GENERAL (Cont'd):

H. Consultant to indicate method of payment.

I. Service authorized by XYZ other than those specifically outlined in the scope of services shall be considered additional services. Additional compensation shall be mutually agreed upon by XYZ and the consultant and shall be invoiced at an hourly rate of $_______ or principal, and $_________ for staff.

J. XYZ shall reimburse the consultant for all out-of-pocket expenses directly chargeable to the project, at actual costs incurred. Such charges shall be itemized and included in the monthly invoices for time charges. Typical reimbursable expenses include travel, lodging, meals, printing costs, long distance telephone calls and other out-of-pocket expenses.

NO GUARANTEE OF THE NUMBER OF UNITS TO BE RELEASED TO THE SUBCONTRACTOR IS MADE OR IMPLIED. AUTHORIZATION TO PROCEED WITH WORK UNDER THIS CONTRACT ADDENDUM WILL BE FORMALLY ISSUED BY THE DIRECTOR OF OPERATIONS IN THE FORM OF A WRITTEN RELEASE OF UNITS FOR CONSTRUCTION. PROFESSIONAL SERVICES WOULD BE RELEASED BY THE DIRECTOR OF PROJECT DEVELOPMENT.

CONTRACTOR:__________________ XYZ CORPORATION__________________
BY:__________________________ BY: __________________________
TITLE:____________ DATE:_____ TITLE:________________DATE:_______

PREPARED BY:____________________________________ DATE:_______
REVIEWED BY:____________________________________ DATE:_______

Appendix 8-5

ARCHITECTURAL WORKING DRAWINGS

CONTRACT ADDENDUM

XYZ CORPORATION ADDENDUM #______________
STREET ADDRESS DATE:______________
CITY, STATE, ZIP PAGE:_______ OF __________
FORMING PART OF CONTRACT # __________________ DATED:__________
BETWEEN XYZ AND______________________________________
FOR WORK AT______________________________________

CATEGORY OF WORK: ARCHITECTURAL WORKING DRAWINGS

NOTE: THE STATED PRICE BELOW INCLUDES ALL FEDERAL, STATE AND LOCAL TAXES.

<u>DESCRIPTION & SCOPE OF WORK:</u>

The consultant named above shall perform the following tasks for XYZ. This scope of services limits itself to the production of construction documents (including an architectural site plan) appropriately signed and sealed.

The above will have the following housing product:______________________________

All of the above shall be in accordance with all of the requirements of __________ County, in respect to the regulations of all the local government agencies. These shall be the Building Department and the Planning and Zoning Department. Their building, fire and zoning regulations shall be properly reflected on the construction documents as well as compliance with their adopted county and/or national codes. The consultant shall inform XYZ when conflicts should occur between the regulations from these local government departments and the original XYZ design, and act in an advisory role.

CONTRACTOR:________________ XYZ CORPORATION________________
BY:________________________ BY: ________________________
TITLE:____________ DATE:_______ TITLE:______________DATE:______

PREPARD BY:______________________________________ DATE:_______
REVIEWED BY:_____________________________________ DATE:_______

CONTRACT ADDENDUM

XYZ CORPORATION ADDENDUM #____________
STREET ADDRESS DATE:____________
CITY, STATE, ZIP PAGE:______ OF ______
FORMING PART OF CONTRACT # ____________ DATED:________
BETWEEN XYZ AND____________________
FOR WORK AT____________________

CATEGORY OF WORK: ARCHITECTURAL WORKING DRAWINGS

I. Construction Documents

A. Prior to starting the work, the consultant shall meet with the Department of Project Development to formalize all the different design and building parameters required for the execution of the construction of this project.

B. The consultant shall finalize the construction drawings including architectural, structural, HVAC, electrical and plumbing designs to sufficient level for building permits, negotiating contracts and construction, including the compliance with energy code. Specifications shall be provided on drawings.

C. All dimensions shall be true and correct. If the XYZ Construction Director deems that the construction details are insufficient for the proper execution of the project, the consultant shall provide the necessary clarifications, including sketches, if necessary, in a timely manner, at no extra charge to XYZ.

D. The consultant shall assist XYZ at any local government meeting, in any interpretation of the construction documents, if required.

E. At submissions to __________ County staff, its Building Department, or its Planning and Zoning Department should there be indicated revisions for non-compliance with the governing building code or any other local regulation concerning planning, zoning or methods of construction, these shall be corrected by the consultant at no extra charge to XYZ.

CONTRACTOR:____________ XYZ CORPORATION____________
BY:____________ BY: ____________
TITLE:________ DATE:______ TITLE:________ DATE:______

PREPARED BY:____________ DATE:______
REVIEWED BY:____________ DATE:______

CONTRACT ADDENDUM

XYZ CORPORATION ADDENDUM #________________
STREET ADDRESS DATE:________________
CITY, STATE, ZIP PAGE:________ OF ________
FORMING PART OF CONTRACT # ____________________ DATED:__________
BETWEEN XYZ AND____
FOR WORK AT____

CATEGORY OF WORK: ARCHITECTURAL WORKING DRAWINGS

I. Construction Documents (Cont'd)

F. Reviews of progress prints shall be done so between the consultant and XYZ Chief Architect, for proper coordination and compliance of the project with XYZ's goals and objectives.

G. The consultant agrees to comply with the set schedule of completion of the work necessary to file for a building permit with _____________ County. This shall be set by the Department of Project Development.

II. Construction Observation

The consultant shall provide on-site construction observation on an as required basis.

III. Fees

A. For professional services to be rendered for the above and bases on estimated hours to be spent by principal and staff, the consultant shall charge a maximum up-set fee of:

_______________.

B. Construction Phase -- Shop Drawings $________________

C. Construction Observation -- Site Visits $________________

These shall reflect the fees per visit, and these shall be as stated before : "On an as requested basis" (reimbursables shall be paid extra).

CONTRACTOR:________________ XYZ CORPORATION________________
BY:____________________________ BY: ____________________________
TITLE:______________ DATE:________ TITLE:________________DATE:_____

PREPARED BY:____DATE:______
REVIEWED BY:____DATE:______

CONTRACT ADDENDUM

XYZ CORPORATION ADDENDUM #______________
STREET ADDRESS DATE:____________________
CITY, STATE, ZIP PAGE:______ OF __________
FORMING PART OF CONTRACT # __________________ DATED:____________
BETWEEN XYZ AND____
FOR WORK AT____

CATEGORY OF WORK: ARCHITECTURAL WORKING DRAWINGS

GENERAL:

A. There shall not be any repeat fees involved in this contract.

B. XYZ reserves the right to reuse the construction documents in any other site or community without having to give any additional compensation to the consultant.

C. XYZ reserves the right of ownership of the construction documents, as defined in Paragraph #2 of the Professional Service Agreement signed between the consultant and XYZ.

D. The consultant shall not subcontract out any work outlined in this agreement without prior approval by XYZ.

E. The consultant will assist XYZ, through attendance at meetings with regulatory agencies in preparation of the construction documents, if required.

F. XYZ reserves the right to stop the work at any moment and remunerate according to the completion of the work up to that moment.

G. Compensation - XYZ agrees to compensate the consultant for the tasks outlined in the scope of services included herein. This shall be paid by XYZ upon approval of the construction documents and condominium exhibits.

H. Consultant to indicate method of payment.

CONTRACTOR:_________________ XYZ CORPORATION__________________
BY:__________________________ BY: ____________________________
TITLE:_____________ DATE:______ TITLE:________________DATE:_______

PREPARED BY:____ DATE:_______
REVIEWED BY:____ DATE:_______

CONTRACT ADDENDUM

XYZ CORPORATION ADDENDUM #____________
STREET ADDRESS DATE:____________
CITY, STATE, ZIP PAGE:______ OF ______
FORMING PART OF CONTRACT # ____________ DATED:________
BETWEEN XYZ AND____________________
FOR WORK AT____________________

CATEGORY OF WORK: ARCHITECTURAL WORKING DRAWINGS

GENERAL: (Cont'd)

I. Service authorized by XYZ other than those specifically outlined in the scope of services shall be considered additional services. Additional compensation shall be mutually agreed upon by XYZ and the consultant and shall be invoiced at an hourly rate of $__________ for principal and $__________ for staff.

J. XYZ shall reimburse the consultant for all out-of-pocket expenses directly chargeable to the project, at actual costs incurred. Such charges shall be itemized and included in the monthly invoices for time charges. Typical reimbursable expenses include travel, lodging, meals, printing costs, long distance telephone calls and other out-of-pocket expenses.

NO GUARANTEE OF THE NUMBER OF UNITS TO BE RELEASED TO THE SUBCONTRACTOR IS MADE OR IMPLIED. AUTHORIZATION TO PROCEED WITH WORK UNDER THIS CONTRACT ADDENDUM WILL BE FORMALLY ISSUED BY THE DIRECTOR OF OPERATIONS IN THE FORM OF A WRITTEN RELEASE OF UNITS FOR CONSTRUCTION. PROFESSIONAL SERVICES WOULD BE RELEASED BY THE DIRECTOR OF PROJECT DEVELOPMENT.

CONTRACTOR:____________ XYZ CORPORATION____________
BY:____________ BY: ____________
TITLE:________ DATE:______ TITLE:________ DATE:______

PREPARED BY:____________________ DATE:______
REVIEWED BY:____________________ DATE:______

Appendix 8-6

LAND DEVELOPMENT SURVEYING
SCOPE OF SERVICES

LAND DEVELOPMENT

Boundary Survey

Includes a research of public land records; location of existing monuments, pipes, markers, etc.; traverse establishment; boundary corner establishment; furnishing of a certified plat which shows property corners and boundary lines with noted bearings and distances as well as any existing man-made features and any other data necessitated by legal and/or public agency requirements; and, the furnishing of a written boundary description.

Topographic Survey

Includes field work and/or photogrammetry necessary to prepare and furnish a contour map of the tract at a scale of 1in. = 50 ft or larger and at a contour interval of 1ft. Indicate rock outcroppings, wooded areas, streams, obvious fill areas, buildings, fence lines and any other existing features.

Land Development Surveying

1. Control

The Surveyor shall be responsible for establishing and bringing forward all horizontal and vertical control required.

2. Earthwork Staking

The Engineer shall provide one (1) complete set of stakes for mass grading the site. A complete set of stakes is considered to be road grade stakes reflecting the centerline road grade and alignment and two (2) stakes to define "house pad" for purposes of mass grading. It is assumed that the original set of stakes will be preserved and protected during the initial phase of mass grading, however, replacement of a maximum of ten percent (10%) of the total number of mass grading stakes originally set will be replaced by the Surveyor at no additional cost to the Owner.

3. Utility Staking

A. Storm Sewer

One (1) complete set of stakes shall be provided on an off-set determined by your Contractor. These stakes shall reflect alignment and grade and will be provided with 50 foot centers along all lines. In addition, all structures will be staked for location and grade. Cut sheets will be provided not later than the working day after the completion of the staking. Grades will be marked on the stakes provided as a part of the staking procedures.

LAND DEVELOPMENT (Cont'd)

B. Sanitary Sewer

One (1) complete set of stakes shall be provided on the off-set determined by your Contractor. These stakes shall reflect alignment and grade and shall be provided with 50 foot centers along all lines. In addition, all structures shall be staked for location and grade. House connections and clean outs (Y's, T's etc) shall be staked with a lath showing the location of the plan station. Cut sheets will be provided.

C. Water

One (1) complete set of stakes shall be provided on an off-set determined by your Contractor. These stakes shall reflect alignment and grade and will be provided with 50 foot centers along tangents and 25 foot centers along curves. Stakes shall also be provided for all tees, valves and fire hydrants. Curb line shall be staked as required sufficient to setback curb boxes. Cut sheets will be provided with grades to top of curb.

D. Electric/Telephone Crossings

Underground electric and telephone crossings of the roadways will be staked in accordance with the plans prepared by the electric and telephone companies. These plans will be provided to the Engineer by the Owner.

E. Other Utilities

It is anticipated that all other utilities will be installed without the need for field surveying.

4. Curb Staking

The Surveyor shall set one (1) complete set of alignment and grade stakes at an off-set determined by your Contractor for all curb construction. Stakes will be placed on 50 ft centers along tangents and 25 ft centers on horizontal and vertical curves. On curb returns at intersections, the P.C., P.T. and center of curve will be staked for purposes of assisting the Contractor during the construction of this item. Cut sheets will be provided.

5. Monuments

Monuments will be set in accordance with the final plat pursuant to schedules agreed to with the Owner.

6. As-Builts

The Surveyor, after completion of sanitary, storm drain, water main and fine grading will provide As-Built Plans for each section. The "As-Built" plans will be prepared in accordance with the requirements of the local ordinances.

Appendix 8-7

HOUSELINE SURVEYING
SCOPE OF SERVICES

HOUSELINE SURVEYING

1. Orientation to Outbounds:

This fee will include verifying horizontal and control to existing outbounds, to determine whether or not they are in agreement with the survey prepared. If we should find any discrepancy in the survey, we will notify your office immediately, so that you may contact the Subcontractor responsible for the survey. It will also be the developers' responsibility to provide us with a copy of the most recent survey.

2. Individual Lot Closure:

Includes verification of all metes and bounds illustrated on the Record Plan by others provided by your office.

3. Rough Grade Stakes:

Including the setting of grade stakes on lots to adequately control grading operations.

4. Lot Clearing Stakes:

Including the staking of limits of clearing at rear of houses

5. Utility Stakeout:

Including the staking of the well and septic tank locations in accordance with permit drawings (by others).

6. Driveway Stakes:

Including setting stakes at centerline of driveways by 4 ft offset to curbline

7. Individual Lot Corners:

Includes the setting of iron pins on all corners (or permanent marks in the sidewalk on the property lines extended where applicable) on each lot after fine grading is completed. State Law requires that all lot corners shall be marked unless we have written instructions from the "ultimate user", requesting that they not be set. The "ultimate user" is defined as the purchaser of the property or the attorney representing the purchaser. In the event that the developer does not wish to provide corner markers on all lots covered by this agreement, a statement must be signed by each "ultimate user" prior to the release of the final inspection survey for settlement and/or certificate of occupancy and forwarded to our office specifying that request corners, the developer will be charged $__________ / Lot which will be billed as an additional item in place of the per lot cost listed. The developer should allow for this additional cost to be included with any settlement costs.

All lot corners to be paid by Developer.
Individual lot corners as requested by "Ultimate User" to be paid by Developer.

HOUSELINE SURVEYING (Cont'd)

8. Individual Plot Plan:
Including preparation of plan to a scale of 1in. - 30 ft.
To be used as a base for building permit plan, foundation survey and final survey.

9. Individual Building Permit Plan:
Including location of individual house on plan to be used for application for building permit. No house grades will be set.

10. House and Lot Grades:
Including location of individual houses on lots, setting floor elevations and illustrating same on individual Plot Plans.

11. House Stakes:
Including setting up of four stakes for line and grade to control individual house construction. This item will be applicable on wooded lots where lot clearing or house clearing stakes were previously completed. If lot is not cleared, billing will be on our normal time charge method of billing.

12. Foundation Location Survey:
Including preparation of Title Survey Plan to a scale of 1in. = 50 ft illustrating location of existing house foundation.

13. Final Inspection Survey:
Title Survey Plan to show complete house and type. Including a block and lot description typed on the plot plan. If a metes and bounds description is required, it will cost an additional $__________ / per lot. As-built lot grades are included as per Township requirements.

14. Pad Stakes:
Including calculating location and setting four stakes for line and grade on building pad corners.

15. Re-Staking (for budget purposes only): "TIME CHARGE"

16. Extra Work: "TIME CHARGE"
Includes items of work requested but not included in the above contract. A proposal for said work will be submitted upon request.

It is the Developer's responsibility to protect stakes once they are in place. Re-staking will be charged against item #15 and billed on our normal time method of billing.

The Developer will be charged at actual cost for all prints supplied on the project.

The above is proposed as a comprehensive list of professional surveying services commonly required on a specific development. Certain items may be deleted without affecting the other unit charges.

1 in. = 25 mm; 1 ft = 30 cm

Chapter 9

Sustainable Residential Communities

Why does homeownership matter? According to the National Association of Realtors, homeownership matters to people for a variety of reasons:

- Overall quality of life is more satisfying than renting because homeowners feel much more in control of their lives. They are in control of their home, and it is a source of joy.
- Net worth of a homeowner has historically increased over the years, and the result is building long-term wealth.
- Most homeowners enjoy stable housing costs with fixed-rate mortgages, which makes preparing monthly household budgets easier.
- Homeowners can deduct mortgage interest and property taxes on their federal individual income tax return.
- Homeowners do not move as frequently as renters, become more entrenched in neighborhood stability, and reap the benefits of community involvement.
- Children of homeowners are more likely to participate in organized activities, school groups, and community organizations.
- Pride of homeownership further establishes self-worth, because it requires accepting personal responsibilities.

What, How, When, and Where to Build

Know the Past to Understand the Future

The following questions are not new, but the answers may very well be different today. Residential builders must ask these questions as land development engineers strive to deliver a marketable and sustainable project:

- What to build?
- How to build?

- When to build?
- Where to build?

Consider the distribution of housing stock as presented in the US Census Bureau's 2013 Housing Profile (www.census.gov/programs-surveys/ahs/data):

- 21 % of US homes were built between 2000 and 2010,
- 26% of US homes were built between 1980 and 1999,
- 25% of US homes were built between 1960 and 1979,
- 15% of US homes were built between 1940 and 1959,
- 7% of US homes were built between 1920 and 1939, and
- 6% of US homes were built before 1919.

More than 60% of all existing homes will be 40 years or older in 2020!

The following breakdown represents the annual demand for new housing units (in millions):

Annual demand	Millions
Household formations	1.30
Net removal of housing units	0.25 (and this will increase)
Trend growth of vacant units	0.05
Placement of mobile homes	– 0.15
Units started but not completed	0.05
Required annual number of housing starts	1,500,000

In the perfect world of data analysis, there is a shortage of new homes being constructed following the 2008 Great Recession and taking into account the anticipated population growth. Existing housing stock is aging, and families are staying in their current homes longer. Many baby boomers are retiring in-place and millennials remain on the fence regarding purchasing a home.

What to Build

Builders have included too many bells and whistles as standard amenities, and over time, buyers have become unwilling to accept less.

However, first-time home buyers will not be able to afford the new standards. With price points increasing and requirements of local governments, the first-time home buyer market will not be substantially met by single-family detached homes.

The phrase used to bring home prices down in the 1980s was "nice but not necessary." There is only so much cost savings in the structure to substantially lower costs. Builders must focus their attention on land prices, infrastructure costs, and increased densities. The housing industry always relied on the market to dictate product, features, and finishes, but design and material use have been elevated as key components necessary to obtain local government approvals.

How to Build

Build with quality.

When to Build

If builders build it, will buyers come? Yes.

Where to Build

Real estate has always been about *location, location, location*. Also, the industry has always been fragmented based on markets, but in the future, the industry will be further defined by market segments. However, for the buyer, it is always *community, community, community*. Affordability is not the least expensive home but rather what a home a buyer can afford on a monthly basis in the community of choice. The typical criteria buyers look for when they are buying a home include

- Top-rated schools;
- Recreation facilities, parks, open space;
- Views of nature;
- Variety of shopping;
- Homes with consistency of architecture;
- Economic stability within the community;
- Public transportation availability;
- Healthcare alternatives;
- Quality jobs nearby;
- Location of the home within the neighborhood and appreciation value;

- Environmental concerns (floodplains, wetlands);
- Amount of crime;
- Amount of noise;
- Landscape features of the property;
- Local business health;
- Cost of energy (natural gas, electricity, oil); and
- Proximity to job.

The home buyer criteria highlighted provide insight on sustainable housing projects. Most think of sustainability as either the United Nations definition of "Social–Environmental–Economic" or the rephrasing by the National Association of Home Builders (NAHB), "Balancing the needs of people, planet, profit." Of course, a wider definition of a sustainable project may include the foundation of society and the ecosystem that supports financial, energy, and infrastructure decisions. Regardless of definition, builders will strive to better meet the following sustainable ideals of the home buying public:

- Live: Does the location provide a livable environment for raising a family, becoming active in the neighborhood, and joining a religious community?
- Work: Are there jobs close to where you live? Can you get to those jobs without driving long distances? Is your community thriving with sound economic policies?
- Play: Do you live to work or work to live? Are there recreation areas, open spaces, and cultural opportunities within walking distance or easy access by car?
- Shop: Is a variety of shopping within easy access? Is there a mix of uses in one place so that "one-stop shopping" is achieved?
- Learn: Are there education opportunities for all age groups? Technology will force society toward continuous learning online, but children will continue attending brick and mortar schools, which should be the best a community has to offer.

Sustainability is a reasonable goal but almost impossible for a real estate project to meet. The following smart growth principles may guide our future land development designs:

- Walkable neighborhoods and connectivity between projects;
- Community, stakeholder, and builder partnerships;
- Designed communities promoting a strong sense of place;
- Flexibility in a mix of land uses;

- Preservation of critical environmental areas;
- Transportation choices; and
- Encouragement and promotion of low-impact community and building designs.

Vertical mixed-use projects may be representative of a sustainable project because stakeholders commonly will be included in the decision-making process at the local level. This type of project will keep housing, schools, businesses, and parks within walking distance of each other, reducing dependence on automobiles. Vertical mixed-use projects consolidate areas into residential, commercial, recreational, educational, energy, and transportation components, and mixed uses help manage vehicle use resulting in reduced energy consumption (Figure 9-1).

Figure 9-1. Residential components of vertical and horizontal mixed-use projects remain the foundation of a successful project that meets social, environmental, and economic sustainability. Energy and transportation needs are also an integral part of a sustainable housing project.
Source: David E. Johnson.

Social Sustainability

Mixed use is considered to be an essential approach for projects to meet the definition of social sustainability because they

- Provide places to live, work, shop, and play;
- Reduce impervious surface by allowing parking facilities to be utilized by noncompeting uses at different times of the day;
- Promote pedestrian activity and movement over vehicular; and
- Provide employment opportunities in a mixed-use project with office, retail, and commercial components.

Environmental Sustainability

A low impact development (LID) design approach provides a step toward sustainable projects. The approaches may

- Focus on land conservation and natural open space,
- Limit impervious surface by reducing pavement widths within public rights-of-way by design,
- Use disconnected runoff flow patterns from built-upon areas to the public drainage system, and/or
- Use structural and nonstructural pollution control measures integrated into the infrastructure design.

Economic Sustainability

Economic sustainability in any community is difficult to achieve and measure, because it is influenced by market, physical location and constraints, regulatory climate, financial stability, and political issues. Economic sustainability is constantly changing because of growth pressures from all real estate sectors.

A mixed-use project is structured to provide jobs and an economic base for a community.

Communities are asking for cost benefit analysis reports, which further substantiates that mixed-use projects are economically sustainable.

Did You Know?

Smart Growth Network

EPA, along with nonprofit organizations, other federal agencies, trade organizations, and professional groups, formed "Smart Growth Network" in 1996. The Smart Growth Network works to encourage development that boosts the economy, enhances community vitality and protects the environment through its network of partners.

Livable Communities

The following acts have been introduced in Congress but have all been remanded back to committees for additional work. This does not mean the intent is dead, but they will most likely be modified and reintroduced again and again:

- S. 1619 (111th) Livable Communities Act of 2009, introduced August 6, 2009.
- S. 1621 (112th) Livable Communities Act of 2011, introduced September 22, 2011.
- Livable Communities Act of 2009. In this act, the legislation defines *livable community* as a metropolitan, urban, suburban, rural, or neighborhood community that
    - Provides safe and reliable transportation choices;
    - Provides affordable, energy-efficient, and location-efficient housing choices for people of all ages, incomes, races, and ethnicities;
    - Supports, revitalizes, and encourages the growth of existing communities and maximizes the cost effectiveness of existing infrastructure;
    - Promotes economic development and economic competitiveness;
    - Preserves the environment and natural resources;
    - Protects agricultural land, rural land, and green spaces; and
    - Supports public health and improves the quality of life for residents of and workers in the community.

The acts include the goals and objectives of smart growth. They clearly promote energy-efficient and location-efficient housing choices. There are other well-established organizations with similar initiatives and with federal agencies backing them to swing the land development pendulum.

The following groups and short descriptions are taken directly from their websites.

Sustainable Cities

Intelligent traffic solutions, green buildings, water management, and smart grid infrastructure are just a few of the technologies helping to steer today's urbanization toward sustainability.

For a real-world look at how solutions can be implemented today, the report "Smarter Neighborhoods, Smarter City" contains detailed recommendations on how to help America's largest urban area—the city of New York—plan for more sustainable growth (www.urenio.org/wp-content/uploads/2012/04/smarter-neighborhoods-smarter-city). Intelligent traffic solutions, green buildings, water management, and smart grid infrastructure are just a few of the technologies helping to steer today's urbanization toward sustainability (www.hud.gov/program_offices/economic_development/Six_Livability_Principles).

HUD, DOT, and EPA

On June 16, 2009, the US Department of Housing and Urban Development (HUD), DOT, and the EPA joined together to help communities nationwide improve access to affordable housing, increase transportation options, and lower transportation costs while protecting the environment. The partnership agencies incorporate six principles of livability (www.hud.gov/program_offices/economic_development/Six_Livability_Principles) into federal funding programs, policies, and future legislative proposals to

- Provide more transportation choices:
  Develop safe, reliable, and economical transportation choices to decrease household transportation costs, reduce our nation's dependence on foreign oil, improve air quality, reduce greenhouse gas emissions, and promote public health;
- Promote equitable, affordable housing:
  Expand location- and energy-efficient housing choices for people of all ages, incomes, races, and ethnicities to increase mobility and lower the combined cost of housing and transportation;
- Enhance economic competitiveness:
  Improve economic competitiveness through reliable and timely access to employment centers, educational opportunities, services,

and other basic needs by workers, as well as expanded business access to markets;

- Support existing communities;
  Target federal funding toward existing communities—through strategies such as transit-oriented, mixed-use development and land recycling—to increase community revitalization and the efficiency of public works investments and safeguard rural landscapes;
- Coordinate and leverage federal policies and investment;
  Align federal policies and funding to remove barriers to collaboration, leverage funding, and increase the accountability and effectiveness of all levels of government to plan for future growth, including making smart energy choices such as locally generated renewable energy;
- Value communities and neighborhoods; and
  Enhance the unique characteristics of all communities by investing in healthy, safe, and walkable neighborhoods—rural, urban, or suburban.

HUD, DOT, and EPA are implementing the Livable Communities Act of 2009 (https://www.congress.gov/bill/111th-congress/senate-bill/1619?s=1&2=9) without the legislation being passed by Congress. These agencies devote dollars and a labor force to promote livability principles at the local level.

National League of Cities: Sustainable Cities Institute

Sustainable Cities Institute (https://sci.uoregon.edu) is the National League of Cities' centerpiece initiative to catalyze, inform, celebrate, and support local governments as critical leaders in advancing and strengthening comprehensive approaches to sustainability. The Sustainable Cities Institute aims to guide and inform local governments that want to pursue sustainability—in their own operations, as well as across the communities where they govern. This commitment to sustainability will challenge local leaders to find the best long-term solutions for balancing environment, equity, and economy. To implement these solutions, local government leaders need to prepare their communities for change.

US Chamber of Commerce's Business Civic Leadership Center

The annual Siemens Sustainable Community Awards program is a partnership between Siemens Corporation and the US Chamber BCLC.

Smart Cities Dive

"Smart Cities Dive provides in-depth journalism and insight into the most impactful news and trends shaping cities and municipalities. The newsletters and website cover topics such as transportation, building and infrastructure, governance, tech and data and more" (https://www.smartcitiesdive.com).

HUD

The mission of HUD's Office of Sustainable Housing and Communities is to create strong, sustainable communities by connecting housing to jobs, fostering local innovation, and helping to build a clean energy economy.

Center for Sustainable Organizations (CSO)

CSO is a 501(c)(3) nonprofit corporation created in 2004 (https://www.sustainableorganizations.org).

Sustainable Community Development Institute (SCDI)

Working with extractive and land-development companies, SCDI has developed a successful approach to deliver sustainable social and economic benefits to host communities by advocating integration of community engagement in the core business activity of the company from the early stages of project development through construction, operation, and closure (https://www.scdinstitute.org).

Community Revitalization Alliance

This organization works on behalf of 50 organizations and institutions in 20 states and is hosted and managed by Sustainable Community Development Group, Inc. Groups and the institutions in the alliance are leaders in problem-solving and implementing innovative rural and urban strategies and projects. Its work epitomizes place-based best practices (http://www.sustainablecommunitydevelopmentgroup.org/revive_alliance.htm).

Sustainable Community Initiatives

Its mission is to promote and develop collaborative community projects and public education programs that enhance economic opportunities and connect businesses, institutions, and people more fully to their communities, the natural environment, and to each other (http://www.sustainablecommunityinitiatives.org).

Smart Growth Rating Systems

The list of groups and their initiatives continues and continues. The list of smart growth rating systems also continues to expand and includes

LEED: Leadership in Energy and Environmental Design (https://www.usgbc.org/leed)

GreenLITES: Green Leadership in Transportation Environmental Sustainability (https://sustainable-infrastructure-tools.org/tools/greenlites)

Green Streets (https://www.epa.gov/G3/learn-about-green-streets)

Green Highways Partnership (https://www.fhwa.dot.gov/publications/focus/06jun/04.cfm)

STARS: Sustainability Tracking, Assessment and Rating System (https://stars.aashe.org)

Additional Sustainability Resources

Other resources beneficial for learning more about sustainable communities include

American Planning Association's Sustainable Communities Division (https://www.planning.org/divisions/sustainable)

ASCE Committee on Sustainability (https://www.asce.org/templates/membership-communities-committee-detail.aspx?committeeid=000009352852)

FEMACenter for Sustainable Communities (https://www.fema.gov/emergency-managers/national-preparedness/mission-core-capabilities)

However, the typical stand-alone residential project may include some elements of sustainability, but until the market and local governments embrace multiple uses on the same property, the ability for residential land development projects to meet sustainable criteria will be difficult to achieve.

Chapter 10

Conceptual Design Planning

The conceptual design planning phase of a project has always been the most important aspect of project development. This phase establishes the foundation for the entire project design; it also determines the ultimate cost of the project. The conceptual design will delineate public and privately owned properties. Properties within a residential project are commonly controlled by covenants, conditions, and restrictions (CCRs) and recorded as part of the HOA documents. In the conceptual design, stormwater management areas, parks, and dedicated open space areas are typical properties controlled by the HOA.

Conceptual design is the creative part of the land development process. This creativity, however, must be tempered with knowledge of both financial limitations and the housing product to be marketed on the site. Even when a land plan has been prepared and contains the fine attributes of a community design, still the developer must ask, "Can the project be built to sell homes or lots and make a profit?"

Builder/developers should hire consultants for the land planning phase, because the value received will far exceed the cost of the consulting services. The cost associated with preparing alternative land plans to evaluate gross lot yields and land development costs is a prudent business expenditure. This land planning approach should be considered before the purchase of land for development. The conceptual land plan can be used during negotiations with a seller, because it reveals the ultimate lot yield of a site. The lot yield is the ultimate indicator of the land value. Designers should use the due diligence report as a guide in preparing land planning options (see Chapter 4). Illustrating buffers, protective land areas, wetlands, floodplains, and so forth on the land plan are essential for a well-conceptualized plan.

Conceptual design planning has four parts: Preparing the base map, preparing the conceptual land plan, selecting a housing product, and preparing a conceptual cost estimate.

Base Map

A base map for land planning is prepared from a boundary survey and a topographic survey (Figure 10-1). Appendix 10-1 presents a detailed checklist for the elements of a boundary survey. To ensure the survey is complete, review the plan prepared by the surveyor and then use the checklist to note items that appear on the plan. The base map must contain all of this information before you can undertake a meaningful land plan.

Appendix 10-2 presents a checklist for the topographic survey work, which, combined with the boundary survey, provides a complete base map for land planning. The topographic information can be obtained through aerial photography (Figure 10-2). On sites of 10 acres or less, a

Figure 10-1. Base map. An accurate topographic and boundary property survey is necessary for preparing the project base map. The base map is the foundation for the project land plan and infrastructure design.
Source: Sidney T. Burns. Due Diligence Associates, Parkland, Florida.

Figure 10-2. Aerial views. An aerial view of the property is a valuable tool for land planning purposes. Assessing the relationship of the property with adjacent land uses and existing infrastructure is concluded using an aerial photo.
Source: Sidney T. Burns. Due Diligence Associates, Parkland, Florida.

field topographic survey can be done to produce a more accurate base map for the project site.

Conceptual Land Plan

The base map is prepared by a surveyor and includes the boundary and topographic survey information. In preparing the conceptual land plan, the land planner analyzes the site conditions and constraints to determine the lot yield. The conceptual land plan should record existing roads and access points, topography, drainage, environmental considerations, soils, vegetation, and utilities (Figure 10-3).

Figure 10-3. Conceptual land plan. Existing site views and site attributes, such as an existing water feature, need to be incorporated in the conceptual land plan.
Source: Sidney T. Burns. Due Diligence Associates, Parkland, Florida.

Roads and Access Points

The land plan should show the existing frontage road and elevations; this helps to determine the best location for the subdivision to intersect with the perimeter roadway. The proposed access location should be evaluated against the existing site constraints, such as telephone poles, fire hydrants, and so on. Select the location of the road access to eliminate any conflict with existing facilities. The vertical and horizontal site distance requirements at the proposed intersection should be evaluated. As an example, the access to the site should not be located on a horizontal or vertical curve that restricts the sight distance at the intersection. Project costs increase if the perimeter existing roadway must be improved as a result of the project. These improvements may consist of road widening, intersection improvements, turning lanes or signalization. Including this information on the land plan will assist the developer in assessing the cost ramifications of the project.

Topography

Review the information on the base map. The closer together the topographic lines, the steeper the slope. The typical slope for residential streets and driveways is no greater than 10% grade. Grades of 10% or more may require a different housing design if the existing topography is to be used. Steep slopes may require excessive cuts and fills that increase the cost of land development.

Wider distances between topographic lines indicate a flatter slope. Thus, by reading the undulations of the site on a topography plan, you can determine whether the site requires special engineering techniques to maximize lot yield and minimize land development costs.

Drainage

Natural drainage swales or channels within the site must be clearly shown on the land plan. Natural drainage courses are depicted by topographic lines that are not straight but curved. These natural drainage areas should be set aside in the land plan for potential drainage conveyance. If there are several drainage areas within the site, evaluate each one in the land plan to determine the most cost-effective design. Altering the natural drainage areas on site not only increases per-lot costs, but it also will not be well-received by the regulatory reviewer.

Environment

The land plan must show wetland areas, wetland buffers, and 100-year floodplain areas. Bear in mind that land set aside for environmental protection reduces the flexibility of the land plan and the number of potential lots within the subdivision. The net buildable area establishes the lot yield and lot costs for land development activities.

Soils

Investigate the site's existing soil characteristics because the type of soil affects the way the project is developed. Most sites have two or more soil types. The land plan should recognize the soil types and use this information to the benefit of the project. As an example, soil types with excellent infiltration capabilities should be used for septic systems and/or drainage basin designs. High groundwater and unsuitable soil, rock, clay, or soil excavated for a profit should be fully addressed during the land plan and engineering design phases.

Vegetation

Existing trees that should be preserved because of the species, size, or characteristics should be shown on the base map, and the land plan should attempt to use this site characteristic to the benefit of the project. The replacement of mature landscaping with similar material is cost prohibitive. However, the market potential for saving specimen trees on site or transplanting the trees on site with the use of a tree spade can be substantial and certainly outweighs the cost of replanting.

Utilities

The locations of all utilities that serve the project site should be indicated on the land plan:

- Sewer lines, including any manhole connections, inverts, and rim elevations;
- Water lines, including the size, type, and class of pipe;
- Fire hydrants and valves;
- Gas mains;
- Overhead electric lines and telephone poles; and
- Drainage culverts under existing roadways or street inlets and connecting storm drains.

Housing Product

Once the conceptual land plan is complete, the land planner meets with the builder/developer to determine the housing product planned for the site (Figure 10-4). To market single-family subdivisions successfully, a builder/developer must find ways to show how the project differs from the competition. The site analysis and conceptual land planning phases have identified the site characteristics and constraints that will govern the design. If used properly, the site attributes will be maximized to their greatest market potential and the resulting housing product will

Figure 10-4. Housing product. Based on the land use and project design program prepared by the builder, the housing product design and architectural elements are consistent with the market and site characteristics.
Source: David E. Johnson.

complement the site characteristics. Thus, the marketability of the homes in the subdivision is enhanced and absorption will be increased.

Housing types that complement the terrain and the site characteristics help to minimize land development costs. If the land and land development costs are lower than those of the competition, the homes can be offered at a greater value. Home buyers understand value—if the project demonstrates value, they will purchase the product regardless of whether the value is real or perceived. If the difference in land and land development costs is always taken as profit, then there will not be any significant difference between competitive projects. Wise developers succeed by sharing the profit with value and site uniqueness.

Foundations

As part of selecting the housing type, the builder/developer considers the various types of foundations and selects one appropriate for the marketplace and the site.

- Basements. Basements are ideal for generating fill and adding to the value of the home. They have an excellent perceived value and, if the site topography has dramatic changes, basement homes provide flexibility in grading and clearing work.
- Side-to-side and front-to-back splits. This housing type uses the ground to establish greater market appeal, and the cost savings can be used for added values.
- Crawl spaces and basements. These features allow minimal clearing of lots for housing construction. The minimum distance from the finished floor elevation to the finished grade should be no less than 1.7 ft.
- Slab construction. These foundations are excellent for flat areas. Pads for slab foundations must be created during earthwork operations. The minimum distance from the finished floor elevation to the finished grade should be no less than 0.7 ft.

The builder/developer may consider other housing styles for areas where the terrain varies in elevation, such as garages under the home, split levels, walkout basements, and others. The housing design and land plan are complementing elements that, if approached with practical market and design sense, will increase sales and provide customer referrals.

The capability to develop and build the project in phases reduces cash flow requirements and regulatory commitments, so proper project phasing is another part of preparing the conceptual design plan. The object is to provide as many lots as possible for the least amount of infrastructure expense. Not only is this approach advantageous for cash flow purposes, but it helps to manage customer lot selection. Each of the project phases should include a mix of quality lots, with the best-located lots kept aside for the later phases; these lots will always sell. Establishing premiums on quality lots also helps to sell the less-attractive lots first.

Covenants, Conditions, and Restrictions

If the business plan includes selling lots to outside builders or to individual lot purchasers, architectural covenants and restrictions should be developed during the conceptual land plan phase. The housing product selected for the project establishes a baseline for meeting the market. The basic elements of CCRs follow (Figure 10-5).

Description of Permitted Uses

This statement is based on the zoning regulations imposed by local government. It can be expanded to include specific accessory uses and buildings such as sheds, detached garages, or other types of buildings that could be constructed on the lot. The description of the permitted uses controls the aesthetic appearance of each lot. Controlling the street scene maintains property values and supports the sales effort for lots or homes in the subdivision.

Figure 10-5. Covenants, conditions, and restrictions. Developers will prepare a document controlling the design features of homes constructed within the subdivision. The intent is to protect the property values of the project when multiple builders are constructing homes within the project.
Source: David E. Johnson.

Type of Building

If the housing product is a ranch or a two-story colonial style and the preference is to have that type of product built in the subdivision, the building restrictions should be included in the CCRs. The CCRs can be more restrictive than local zoning and planning ordinances. Typically, the type of building is judged on the style and size of the house; for example, the only houses permitted would have to be at least 2,000 ft^2 two-story colonial contemporaries with a two-car garage. These types of building restrictions can be included in the CCRs to govern the future buildout of the project site.

Height Restrictions

Height restrictions can be imposed in the CCRs by limiting the style of the house to one, two, or three stories. This restriction also sets the project apart from the competition. The height restrictions must meet the bulk requirements of the zone.

Setbacks

In general, setbacks from the street, side yards, or rear yard complement local zoning bulk requirements but can be more restrictive for accessory buildings and garages.

Architectural Restrictions

Architectural restrictions are imposed to maintain the aesthetic appearance of the street scene. The restrictions could include a look-alike provision, which prohibits similar exterior housing elevations from being constructed adjacent to or across the street from each other. Particular attention should be made to the roofline, window locations, colors, and door styles. Façade materials such as vinyl, wood, brick, or stone, and the quality of the material, could be dictated in the CCRs to control the house types in a subdivision.

Architectural restrictions can be expanded to include exterior colors of housing products, such as shingles, siding, garage doors, front doors, shutters, and so on. Preapproved types of windows, front doors, and garage door materials can be included as well. For instance, the use of steel or aluminum garage doors, paneled or flat, with or without sunbursts would be included in the CCRs.

When the information in the CCRs is vague, its interpretation is open to each individual owner or builder. When restrictions are imposed and the effort is extended to manage the CCR provisions, the author of the CCR must develop a guideline that is very specific in terms of criteria and standards to be followed by each lot owner. A sample outline of a CCR document, which can be used for single-family subdivisions, is provided in Appendix 10-3. A sample CCR for a single-family home is included in Appendix 10-4.

Another aspect of implementing CCRs is the single requirement that all proposed house plans be approved by the developer. The developer reviews the house plans, the material specifications, and the construction specifications before any building permits are issued. This benchmark is inserted into CCRs to ensure that all homes meet the standards for the project. Also in the CCRs, a variance or an appeal process must be provided to give each individual lot owner an opportunity to meet or strive to meet the intent of the CCRs.

CCRs are imposed to maintain property values and control individual taste. The results of implementing this type of control will be consistency in housing styles. CCRs review the individual design approaches and compare them to the subdivision as a whole for a homogeneous but individualistic subdivision appearance.

At the completion of a residential housing project, the infrastructure improvements must be transferred by the builder/developer to another entity, such as an HOA. It is in the developer's best interest to turn improvements to the municipality or HOA as soon as it is practical; Appendix 10-5 outlines a comprehensive approach for doing so. This approach is applicable for planned unit developments, small clustered subdivisions, and traditional subdivisions with common elements, such as stormwater management facilities. The outline helps to ensure that all issues are adequately addressed, so the entity required to accept the improvements will not be able to find a reason not to accept the facilities. Thus, the reputation of the developer is further enhanced.

Conceptual Cost Estimate

The last step of conceptual land planning is the preparation of a conceptual land development cost estimate. Preparing an estimate without further design information is appropriate to ascertain the viability of the conceptual plan. A conceptual cost estimate also provides the opportunity to identify all of the land development costs before proceeding with the project. Estimating the finished lot cost based on the conceptual land plan provides an excellent way to assess the lot yield, land development costs, and sales prices.

Appendix 10-6 can be used in developing the conceptual cost estimate for the project. The comments category should be used to indicate how those costs were calculated. This form also provides the means to review the project costs as the preliminary and final designs are prepared. In Appendix 10-5, in the absence of specific design detail, use the rule of thumb included after each line item. The resulting cost estimate should be used as a guideline and represents the conservative approach to land development. Include all of the costs in the cash flow analysis or project pro forma. Appendix 10-7 is a typical houseline budget form that can be used for conceptual cost analysis. Calculate the land development costs per unit. If the goals can be achieved by adding these costs to the other project costs and the housing sales price, the proposed land plan works. Appendix 10-8 is a basic project pro forma that highlights the major cost items for a housing project.

The conceptual cost estimate provides the financial framework to determine what the project can afford and what can be compromised during the regulatory process. Refinements of the land plan will improve the marketability of the project. If the costs are substantially higher than those budgeted in the project pro forma, the land plan should be redesigned to reduce infrastructure costs, increase the number of lots, or a combination of both. The builder/developer should consider whether the project's raw land price may be too high to sell houses and make a profit. The value of the property could well be overinflated, as is often the case for owners who have held the property for a long time.

If the conceptual cost estimate and the pro forma price the project out of the market, do not force the issue. Stay with the housing goals and keep searching for the best-suited property for the marketplace. The conceptual land plan and cost estimate are excellent tools for negotiating the land purchase. If the cash flow analysis for the project shows that the property is properly designed for its highest and best use, proceed with serious negotiations for the purchase of the land.

Appendix 10-1

BOUNDARY SURVEY CHECKLIST

1. Caption giving property location (local, legal, land subdivision, municipality, County, State).
2. Point of beginning.
3. Tie to established reference point.
4. Parcel traversed in clockwise direction.
5. Type corner markers found and set indicated.
6. Courses shown in degrees, minutes, and seconds east or west of a north/south line.
7. Traverse closed.
8. Contiguous property owners indicated.
9. Name of streets, railroads, rivers, creeks, etc. shown.
10. Easements of record located.
11. Easements or rights-of-way contiguous to or ending at boundary shown with ownership noted.
12. Location and dimensions adjacent road rights-of-way.
13. Improvements located.
14. Encroachments located or noted, if none exist.
15. Flood plain note.
16. Curve data shown (points at curvature, tangency, and compound curvature; radii; central angle; and length and bearing of long chord).
17. Error of closure
18. Area (SF, if less than one acre and acres, if larger).
19. Metes and bounds description.
20. Easement descriptions.
21. Plat scale (1 in. = 50 ft.).
22. Date of survey and plat.
23. North arrow (note whether true or magnetic).
24. Certification by surveyor.

1 in.= 25 mm; 1 ft = 30 cm; 1 sq ft = 0.9 m^2; 1 ac = 0.4 ha

Appendix 10-2

TOPOGRAPHIC SURVEY CHECKLIST

1. Boundary lines.
2. Topographic data, including 150' beyond property.
3. Letter of explanation, if permission to go on adjacent property not granted.
4. Location of structures, buildings, other man-made objects and important natural features.
5. Building floor elevation and other important elevations.
6. Location, size and type of materials for water lines, valves, meters, fire hydrants and other appurtenances servicing the property.
7. Location, size and type of materials for sanitary sewer lines servicing the property.
8. Sanitary sewer manholes with invert and top elevations
9. Direction of flow - sanitary sewer.
10. Location, size and type of materials for storm sewers and culverts servicing the property.
11. Catch basins and manholes located with invert and top elevations.
12. Storm flow direction for pipes, ditches, etc.
13. Location and size of gas lines servicing the property.
14. Electrical and telephone lines, including pole locations, servicing, located on or adjacent to the property.
15. Identity of utility owners and operators.
16. Outline perimeter of heavily wooded areas.
17. Location and site of solitary trees with 6 in. and larger diameter trunk.
18. Bench marks clearly described, one per four (4) acres with elevation given to 100th of foot.
19. Contour interval - one foot or as specified.
20. Elevations at high and low points.
21. Elevations at street intersections and every 50 ft along street at edges of pavement (both), curb and sidewalks.
22. Mean elevations and limits of standing or flowing water.
23. Limits of 100-year flood elevation or note, if 100-year flood does not encroach.
24. Identity of flood information source.
25. Location and ground elevation at all soil borings.
26. Caption showing location of property.
27. Map scale - 1 in. = 50 ft, or as specified.
28. North arrow.
29. Date of survey.
30. Certification by surveyor.

1 in. = 25 mm; 1 ft = 30 cm; 1 ac = 0.4 ha

Appendix 10-3

DECLARATION OF COVENANTS, CONDITIONS, AND RESTRICTIONS

SECTION I - DEFINITIONS

1.1	By-Laws
1.2	Charges
1.3	Homeowners
1.4	Declarant
1.5	Dwelling Unit
1.6	Final Plan
1.7	Lot
1.8	Majority of Members
1.9	Member
1.10	Neighborhood
1.11	Owner
1.12	Parcel
1.13	Property
1.14	Residential Owner

SECTION II - APPLICABILITY

2.1	Applicability
2.2	Interpretation of Declaration

SECTION III - MEMBERSHIP

3.1	Membership
3.2	Termination of Membership
3.3	Voting Rights

SECTION IV - COVENANT FOR MAINTENANCE CHARGES

4.1	Creation of the Lien and Personal Obligations of Charges and Assessments
4.2	Owners' Negligence
4.3	Effect of Non-Payment of Charges and Assessments; Remedies of the Community Association
4.4	Lien of Charges
4.5	Annual Charges
4.6	Special Assessments

Appendix 10-4

Examples of Covenants, Conditions, and Restrictions

In order to preserve the character and integrity of a residential community and to protect the value of the homes therein, the declarant hereby declares that the homes and property shall be subject to all CCRs of record and to the following restrictions, all of which shall be perpetual in duration and run with the land:

1. No noxious or offensive activities shall be carried on, in, or upon the property or in any home, nor shall anything be done there either willfully or negligently, which may be or may become an annoyance or nuisance to the other residents in the subdivision.
2. Each home constructed on each lot must consist of no less than 1,750 square feet of living space. The design of said home shall be of a style that is compatible with all other homes within the entire tract and the property. No home shall be constructed, no outbuilding or accessory building shall be constructed, and no addition or alteration shall be made to any existing building without the prior written approval of the declarant. All plans and renderings shall be submitted to the declarant for approval at least 30 days before any applications are filed for a building permit.
3. The declarant shall have a reasonable period of time, but less than 30 days, to review the same. If the plans and renderings are, in the sole discretion of the declarant, contrary to the provisions of this paragraph, the homeowner shall be prohibited from submitting the same for a building permit. The homeowner shall revise or modify the plans and renderings to comply with the design and size requirements of this paragraph and resubmit the same to the declarant. Thereupon the review period shall begin again.

Architectural Review

1. No building, fence, wall, structure, or any improvement shall be erected on the owner's lot, nor shall any exterior change to the color scheme or alteration be made until the plans and specifications showing the nature, height, materials, colors, and location of change have been submitted in writing to the architectural control committee. Such committee shall be established as provided in the Homeowners Association (HOA) bylaws. In the event said committee fails to approve or reject, with or without condition, the application within 30 days after submission, approval will be deemed to have been granted without

conditions. The architectural review committee shall adopt design criteria and standards for architectural control.

2. No owner shall erect or maintain an antenna on any lot.
3. The owner of each dwelling unit or lot shall maintain the building or lot in good order and repair and be consistent with all those CCRs, rules, and regulations as may apply to the building or lots. In the event that a dwelling unit, lot, or parcel shall not be so maintained, the HOA shall have the right to enter the building or lot to maintain the same, after giving the owner at least 20 days' written notice, to cure any maintenance problems. The HOA shall have the right to assess the owner for the cost of such maintenance. The architectural review committee shall establish rules and regulations governing the exterior maintenance of any building or lots.
4. No owner shall carry on any practice, or permit any practice to be carried on, that unreasonably interferes with the quiet enjoyment by any other owner.
5. Each owner of a dwelling unit shall maintain fire and extended coverage insurance.
6. No owner shall leave any nonoperating vehicle or unlicensed vehicle on the lot so as to be visible from any neighboring lot.
7. No owner shall be permitted to lease his or her dwelling unit unless the lease has been submitted in writing to the architectural review committee for approval. All leases shall provide that the lease shall be subject, in all respects, to the provisions of the bylaws, the declaration of CCRs, and rules and regulations of the HOA.
8. No above-ground swimming pools may be erected on any lot.
9. No trailers, vans, storage tanks, temporary structures, or accessory buildings shall be erected on any lot without the approval of the architectural review committee.
10. No recreational vehicles, trucks, or boats shall be parked or stored on any lot.
11. No unsightly weeds, underbrush, or other unsightly vegetation shall be permitted to grow or remain on any lot.

Appendix 10-5

HOMEOWNERS ASSOCIATION CHECKLIST TURNOVER SCHEDULE					
	Community Manager	Legal	Other	Date Start	Date Complete
1. Date released for sale					
2. Documents recorded Copy received					
3. Anticipated closing of first units					
4. Set up contracts					
A. Management					
B. Base contract and addendum approved					
C. Insurance					
D. Trash					
E. Lawn maintenance					
F. Lease agreement approval					
G. Building maintenance					
H. Pool maintenance					
I. Apply for Tax I.D. #					
J. Others					
5. Deposits for:					
A. XYZ					
B. Utility					

	Community Manager	Legal	Other	Date Start	Date Complete
6. Establish bank accounts Signature cards					
A. Special reserves 2 months					
B. Reserve - Monthly					
C. Operating - Operating expense					
7. Set up financial records					
A. Accounts Receivables					
B. Accounts Payables					
8. Set up files					
A. Alpha and numerical Files for owners					
B. Chronological file					
C. Miscellaneous alpha file					
9. After lease agreement approved					
A. Notify real estate agent of procedures					
B. Notify owners					
10. Letter to owners					
A. Notification of Assessment fee payment					
B. Set up coupon payment book - printed					

	Community Manager	Legal	Other	Date Start	Date Complete
11. Set up rental package					
A. Welcome letter					
B. Include rules and Regulations					
12. Set up delinquent letter To be read					
13. Select and appoint Nominating & Transition Committee					
14. Solicit candidates for Board of Directors					
A. Send questionnaire to all owners					
B. Hold meeting to determine candidates					
C. Establish election & turnover date					
D. Notify all departments XYZ					
15. Mail election package certified mail					
A. Notice of meeting					
B. General information					
C. Candidate's résumés					
D. Ballot information					
E. Ballot					
F. ID / Proxy					
G. Agenda					

	Community Manager	Legal	Other	Date Start	Date Complete
16. Preparation of Turnover Documents (per chpt 718)					
A. Declaration of Condominium & Amendments					
B. Articles of Incorporation					
C. Copy of By-Laws					
D. Minute book and Corporate Seal					
E. House rules & regs					
1. Resignation of officers - letters					
2. Financial records					
a. Preliminary Balance Sheet					
b. Advance & Delinquent Payments					
c. Tax Records					
d. Review of records by out-side accountant					
F. Association Funds					
1. Bank Account Balances					
2 Checkbooks.					
G. Inventory of Personal Property					

	Community Manager	Legal	Other	Date Start	Date Complete
H. Plans and Specifications As Builts					
1. Roads					
2. Drainage					
3. Sprinkler system					
I. Insurance Policies					
J. Certificate of Occupancy					
K. Other permits - Swimming Pool					
L. Written Warranties					
M. Roster of Unit Owners					
N. Lease of Common Elements or Other Leases					
O. Employment Contract					
P. Service Contract					
Q. Deed for Common Property					
R. Affidavit of No Liens					
17. Establish transition period for turnover -- Directors-elect work with budget, contracts, rules etc.- with XYZ assistance for 30 days (under contract)					

	Community Manager	Legal	Other	Date Start	Date Complete
18. Conduct Election / Turnover Meeting					
A. Establish Quorum					
B. Call for Nomination From the Floor					
C. Call for All Ballots					
D. Count Ballots - Have Tally Sheets					
E. Announce Directors-Elect					
F. Acknowledge Form for Certification of Votes					
G. Present All Turnover Documents Listed Above					
H. Adjourn Meeting					

Appendix 10-6

CONCEPTUAL COST ESTIMATE

Note: Adjust unit price consistent with the locale of the project.

PROJECT ______________________ LOT __________ BLOCK ________

Location ____

Concept Plan Date __________________ By: ______________________

Number of Units ____________________ Date: _____________________

SUMMARY	PER UNIT CONCEPTUAL COST	COMMENT
Common Area:	________	________
Water System: On-Site	________	________
Off-Site	________	________
Sewer System: On-Site	________	________
Off-Site	________	________
Land Development: On-Site	________	________
Off-Site	________	________
Structure:	________	________
Raw Land:	________	________
TOTAL HARD COSTS	________	________
TOTAL SOFT COSTS	________	________
Overhead	________	________
Per Unit Total Cost:	________	________
AVERAGE SALES PRICE	________	________

* FORMULAS PROVIDED

COST ANALYSIS

LOT ______ BLOCK ______ BY: ____________________

DATE: ____________________

DESCRIPTION OF WORK:	Conceptual Cost Per Unit	Comment
Site Feasibility	________	________________
Survey Fees	________	Houseline/Land Development
Site Planning and Design Development	________	________________
Engineering Fees	________	________________
Architect Fees	________	$2000 - $5000 per model
Landscape Architect Fees	________	________________
Civil Eng. Insp. & Certification Fees	________	________________
Architectural Inspection Fees	________	________________
Field Engineering & Survey Fees	________	________________
Soil Testing	________	________________
TOTAL PROFESSIONAL FEES	________	
Regulatory Processing & Permit Fees	________	Include all agencies
Escrow & Regulatory Inspection Fees	________	________________
Water Connection Fees	________	________________
Sewer Connection Fees	________	________________
Miscellaneous Fees	________	________________
TOTAL REGULATORY FEES	________	

*FORMULAS PROVIDED

LOT ______ BLOCK ______ BY: ____________________

DATE: __________________

DESCRIPTION OF WORK:	Conceptual Cost Per Unit	Comment
Planting Materials (Street Trees, Buffers, etc.)	________	________________
Sod / Seed	________	Right-of-Way & Common Area
Irrigation System	________	________________
Berms	________	________________
Entry Feature	________	________________
Perimeter Walls	________	________________
TOTAL LANDSCAPING COSTS:	________	
Project Utilities, Other	________	________________
Cost for Electric Service	________	________________
Entry Feature Lighting	________	________________
Street Lights	________	________________
Temporary Power	________	________________
Amenity Lighting	________	________________
Telephone Service	________	________________
Cable television	________	________________
Mailboxes	________	________________
Directories	________	________________
Trash Enclosures	________	________________
Wheel Stops	________	________________
Amenities:		
________________	________	________________
________________	________	________________
________________	________	________________

LOT ______ BLOCK ______ BY: ____________________

DATE: ____________________

DESCRIPTION OF WORK:		Conceptual Cost Per Unit	Comment
Water Distribution System:	On-Site	*	
	Off-Site	*	
Sewer Collection System:	On-Site	*	
	Off-Site	*	
OFFSITE LAND DEVELOPMENT			
Clearing		*	
Grading			
Soil Import / Export			
Rock / Tree Removal			
Defective Soil Removal			
Storm Drainage (On-site)			
Structures (Bridges, culverts, etc.)			
Roadways			
Special Paving			
Pavement Markings			
Bike Paths / Street Sidewalks			
Curb & Gutters			
Traffic Signs & Traffic Control			
Street Signs			
Miscellaneous:______________			
______________			
TOTAL OFF-SITE LAND DEVELOPMENT COSTS			

* FORMULAS PROVIDED

LOT ______ BLOCK ______ BY: ____________________

DATE: ____________________

DESCRIPTION OF WORK:	Conceptual Cost Per Unit	Comment
ON-SITE LAND DEVELOPMENT		
Clearing	*	
Grading	*	
Soil Import / Export	*	
Rock Removal		
Defective Soil Removal		Use of Materials Onsite-Refer to Soils Report
Storm Drainage (Off-site)	*	
Structures (Bridges, culverts, etc.)		
Roadways	*	
Special Pavement		For Marketing Purposes
Street Sidewalks	*	
Curbs & Gutters	*	
Traffic Signs / Street Signs	*	
Soil Erosion	*	
Driveway Aprons	*	
Street Trees	*	
TOTAL ON-SITE LAND DEVELOPMENT COSTS	______	
TOTAL LAND DEVELOPMENT COSTS	______	

COMMENTS: ____________________

____

____

* FORMULAS PROVIDED

Formulas for Cost Analysis

Structure	Square Footage Cost x (sq ft + ½ Garage sq ft)
Water Distribution System	
On-Site:	(Total Length of Roads x Pipe Unit Price x 2)+ (No. Of Fire Hydrants x Unit Cost)
Off-Site	Same Formula as for On-site
Sewer Collection System	
On-Site	(Total Length of Roads x Unit Price x 2) + (No. Of Manholes x Unit Cost)
Off-Site	Same Formula as for On-site
Off-Site Land Development	
Clearing	Width of Roadway to be Cleared x Unit Price x 1.3
On-Site Land Development	
Clearing	(Right-of-Way + 20 ft x Length of Roads x Unit Price) + (No. of Lots x Unit Price per Lot)
Grading - Earthwork	Strip Topsoil - Total Area of Clearing x 1 ft x Unit Price General Grading-Total Area of Clearing (sq yds) x Unit Price Fill / Lot - (Per Lot Estimate of cu yd) x Unit Price Rough Grade - Total Area of Clearing x Unit Price Topsoil Replacement - Strip Topsoil Value x 0.4 Final Grade - Rough Grade Value x 0.4 (or $7,000 per ac)
Soil Import / Export	Truckload cu yds x 1.2 x Cost Per Truckload
Storm Drainage	1) Locate inlets at every low point and intersection No. of Inlets x Unit Price 2) Locate inlets approx 250 - 300 ft from road highpoint (Use topo as guide) 3) Start with pipe length of 15 inches and increase diameter For pipe length downstream. Total Length x Unit Price
Roadways	Total Length of Roads x Width x Unit Price
Street Sidewalks	Width of Sidewalk x Length of Road x 2.2 x Unit Price
Curbs & Gutters	Length of Roadways x 2.2 x Unit Price
Traffic/Street Signs	No. of Intersections x 6 x Unit Price
Soil Erosion	Linear Footage of Silt Fence x Unit Price No. of Inlets x Unit Price
Driveway Aprons	No. of Lots x Unit Price
Street Trees	(Length of Roadways ÷ 50) x 2.2 x Unit Price

1 in.= 25 mm; 1 ft = 30 cm; 1 sq ft = 0.9 m^2; 1 sq yd = 0.8 m^2

Appendix 10-7

HOUSELINE BUDGET
LINE ITEMS

DESCRIPTION	
OCHD PERMIT FEE	
ZONING PERMIT	
TREE PERMIT	
FOUNDATION PERMIT	
BUILDING PERMIT FEE	
PRE-ENGINEERING FEE	
SEPTIC DESIGN & PLOT PLAN	
HL SURVEYING	
LOT CLEARING	
STRIP TOPSOIL	
FILL MATERIAL	
EXCAVATE BASEMENT	
EXTERMINATOR	
FOUNDATION	
BACKFILL BASEMENT	
LUMBER	
FRAMING	
ROOF TRUSSES	
EXTERIOR FRONT DOOR	
FYPON / COLUMNS	
WINDOW MATERIAL	
ROOFING	
SIDING	

PLUMBING	
HVAC	
ELECTRIC	
GARAGE DOORS	
INSULATION	
SHEETROCK	
GUTTERS & LEADERS	
SHUTTERS	
INTERIOR STAIRS/ RAILS	
SEPTIC SYSTEM & TANK	
SEWER CONNECTION FEE	
WATER CONNECTION FEE	
WELL	
ROUGH GRADE/ DRIVE/ FLAT	
SERVICE WALKS & PATIOS	
DRIVEWAY PAVING	
INTERIOR TRIM MATERIAL	
INTERIOR TRIM LABOR	
WATER TEST	
SOFTENER	
CERAMIC TILE	
PAINTING	
CABINETS/TOPS	
CABS & TOPS INSTALL	
MIRROR & BATH ACCESS.	
VINYL FLOORING	
CLEANING ROUGH/ FINAL	
CARPETING	

HARDWARE	
RANGE/ HOOD/ DISHWASHER	
FINISH GRADE / RE-TOPSOIL	
PLANTINGS	
SEEDING	
FINAL ENGINEERING FEE	
WARRANTY INSURANCE	
CONSTRUCTION EXTRAS	
WARRANTY COMPANY	
200 - AMP SERVICE	
SECURITY SYSTEM	
FIREPLACE	

Appendix 10-8

PROJECT PROFORMA

MARKET:

Sales Price ___________________

Number of Lots: _________

PER LOT ANALYSIS:

1. ________________
Land purchase price

2. ________________
Soft costs, i.e. design professionals, regulatory fees, application fees

3. ________________
Conceptual land development cost for "buildable" lots - use form

4. ________________
Off-site costs (i.e., water, sewer, road improvements)

5. ________________
Real estate commission to sell new homes including advertising

6. ________________
Houseline costs including all lot costs for a finished lot

7. ________________
Interest and finance costs (land development and/or houseline)

8. ________________
Warranty services, project turnover costs

9. ________________
Project overhead and general administrative

10. ________________
Subtotal

Chapter 11

Residential Subdivision Design

Based on the conceptual land plan, the infrastructure of the subdivision is designed by an engineering firm. Many regulatory requirements and design criteria govern subdivision design work. Various design and construction options are available, and each should be reviewed in detail to select the alternatives that enhance the subdivision's appearance and function while substantially reducing its land development costs. Initial discussions between the builder/developer and the engineering firm should focus on which cost-saving design options should be incorporated into the preliminary design documents.

The options presented in this chapter are valid ways to meet the intent of the regulatory engineering design standards. If an option is acceptable to the local government engineer, then your project can achieve success both financially and aesthetically. If the community does not accept some of these ideas, then it may have more stringent subdivision standards and, most likely, higher housing prices. Standards based on conservative engineering and planning criteria add substantial development costs to residential subdivision design with debatable benefits.

Design and Construction Options

The following design and construction options can guide discussions among the engineering design team before they prepare the preliminary design documents.

Clearing

Limit the amount of clearing necessary to construct the subdivision. If the developer plans to sell lots to builders or individual buyers, the design plans should show the clearing of the rights-of-way and a 10 ft strip of land beyond the rights-of-way lines for the construction of the roads. If the developer is also building on the lots, the plans should show minimum clearing of the lot—no less than 10 ft around a dwelling unit and 20 to 30 ft in the rear yard.

If the practice is allowed by law, burning cleared material on site is cost effective. If burning is not permitted on site, stumps and debris must be trucked to a landfill. Some communities permit developers to bury the stumps in an open-space berm that will not be sold to the homeowners. Several other clearing alternatives result in a cost savings to the project. For instance, trees can be chipped on site. The pile of wood chips is stockpiled in a cleared area out of the way of subsequent infrastructure construction. Later, the chips can be used for ground cover within the tree-save areas. Also, many homeowners will use the chips in their yards. If the chips are used in the subdivision, the contractor can sell the wood chips and reduce the cost for clearing. Stumping the cleared area can result in substantial amounts of debris that must be removed from the jobsite. Renting a stump grinding machine that grinds the stumps into mulch is a cost-effective method for disposal of stumps. This mulch can be mixed with topsoil and used on site for open-space top dressing. Grinding stumps and chipping trees on site reduces or eliminates the cost of trucking the material off site to a landfill.

Having specimen trees staked out within the site by a surveyor is the best option for a residential subdivision. The cost of clearing may increase on a per-lot or per-acre basis because it is considered selective clearing. However, saving trees enhances the subdivision's appearance and adds to the value of the lots. The benefit of selective clearing is greater than the associated cost. This is true even if the trees saved are not specimen trees.

Narrower roads reduce the right-of-way requirements and the amount of clearing. Most residential subdivisions have a 50 ft right-of-way that should be reduced to 2 ft beyond the curbline. Doing so saves trees, reduces costs, and adds land to the properties of tax-paying owners. If the government will not accept narrow pavement widths or rights-of-way, design the subdivision with private roads, which do not have a public right-of-way. The lot lines become the property line of the roadway located at the edge of pavement. A private road system would be owned and maintained by the HOA. Although private roads are usually advantageous, sometimes they are judged as a negative by prospective home buyers, who often view HOAs and the ownership and maintenance of private roadways as an added burden.

All in all, though, the benefit of saving trees and marketing an environmentally sensitive subdivision usually outweighs the perceived burden. In some cases, contractors provide a much lower unit price if they can maximize the use of their equipment. Thus, the engineering firm should consider the clearing work as part of the project phasing to maximize the use of the clearing equipment.

Earthwork

As the engineer, design the project to balance the cuts and fills on the site (Figure 11-1). Try to minimize the costs to importing material and the distances that material must be hauled within the site. Avoid large cuts and fills so that material is not moved more than once within a site. In general, large cuts and fills also require many staging areas for the excavated material before its final destination. Moreover, the land for stockpiling excavated material may require clearing beyond the current phase of lots. Thus, to maximize cash flow at the start of a project, carefully review the

Figure 11-1. Earthwork. Using site contours to minimize cuts and fills is the goal of the engineering design. Contractors use GPS-guided equipment to undertake earthwork operations.
Source: David E. Johnson.

clearing and earthwork requirements for the initial phase. Locate any staging areas in the last phase of the project, if possible. Use pans versus trucks and loaders simply because pans are more efficient and cost effective.

Design the vertical and horizontal subdivision road alignments to complement the topography of the land. Road alignments should follow the terrain by being parallel or perpendicular to the contour lines as much as possible. Also, it is preferable to design the roadways to be cut rather than filled so that the lots are automatically higher than the roadway. This option facilitates proper lot drainage and allows the maximum number of trees to be saved.

A lot's topsoil after clearing and grubbing should be stockpiled in a location that can always be used and does not interfere with subsequent infrastructure development or housing operations. Rock areas, or areas of any unsuitable material identified during the site analysis work, are important to avoid, so use the soil information as a guide during the initial design phase.

The housing foundation design also affects earthwork quantities and operations. Consider the overall architectural design of the subdivision, as well as the type of housing product and the choice of foundation. Grade the lots according to the architectural approach and identify the cost of additional earthwork operations. If individual lots are sold as buildable with the housing pads already in place, decide on a clear definition of *buildable* for the specific housing type.

Sanitary Sewers

Sanitary sewer design is governed by standard regulatory approaches handed down from year to year and from state to state. However, consider whether any of the following ideas could provide an appropriate system at the least cost:

- Design for the minimum-diameter pipe size allowed by regulation. Use 6 in. diameter lines with cleanouts wherever possible, instead of the standard 8 in. line with manholes.
- Align the sewer line along the centerline of the public roadway. Maximize the distance between manholes. Design a minimum distance of 400 ft between manholes and use easements on the lots if necessary. Placing sewer lines on lots inside of horizontal roadway

curves may encumber the property. In this case, an easement within the front setback could be an acceptable and cost-effective approach.

- Use pipe slopes that minimize the trench and manhole depths. If possible, keep all sewer lines within 8 ft of the surface, because this trench depth can be excavated with less expensive equipment. Shallow pipe installations are also easier and quicker.
- Review the soils information to determine rock outcroppings or unsuitable soil locations and use it to design sewer lines or other utilities at the least cost.
- Extend the sewer line in a cul-de-sac far enough to allow access to the last lots by maximizing the length of the lateral line providing service to those lots. This eliminates the need to extend the main trunk line.
- Select the pipe material to reduce costs. Polyvinyl chloride (PVC) pipe for sanitary sewers is acceptable in many municipalities, and it costs substantially less than any other type of pipe material used for sanitary sewer lines.
- Analyze the first cost and maintenance cost of connecting the project to an off-site sewer line with a gravity line or with a pump station or force main system. An alternative to the gravity sewer is the pressure sewer system. This sewer system, which is approved for use by many government agencies, uses grinder (or effluent) pumps at each individual house. Each home uses a 2 in. force main connecting to a 4 in. collector force main in the street. Eventually, the lines connect to an outlet manhole or gravity sewer line. The pressure sewer system is more cost effective than a gravity sewer line design but may meet consumer objections.

Water Distribution Systems

Locate fire hydrants in accordance with local fire department regulations. On the plans, scale a 250 ft radius circle around each fire hydrant. To ensure full fire service coverage in the subdivision, every home should be located within the circular area surrounding the fire hydrants. A linear approach with 500 ft between fire hydrants can also be applied.

Minimize the length of lateral lines to each lot. For single-family houses, use one connection per property line, and then use a Y-connection to the individual homes.

Consider joint trenching with sewer or other utility lines to reduce costs. Coordinate this design option with the other utility line designs, thus ensuring that conflicts with streetlights, transformer locations, or cable television pedestals are minimized. Most utility companies design their structural facilities on common property lines.

Investigate the possibility of using pipe sizes of 4 and 6 in.; minimizing the pipe size can save money although the cost difference between pipe sizes is not significant. Evaluate when pipe sizes can be reduced after the service area has been reduced.

Use PVC material instead of the more-expensive ductile iron pipe (DIP) material. If the local jurisdiction does not permit the use of PVC pipe, provide documentation and information for their review. Longevity, maintenance capabilities, and ease of installation are positive attributes of PVC pipe. Take the opportunity to seek approvals for alternative pipe materials for residential subdivisions.

Drainage Facilities

Minimizing storm drainage costs is basic to a well-visualized land plan that uses the topography to its maximum potential, so it benefits the stormwater management facility and drainage system design for the project (for more on stormwater management, see Chapter 15). Rather than make substantial changes to the land profile, follow the natural contour of the land for drainage swales and retention of stormwater before it flows from the project site. If possible, use natural land forms regardless of their location. However, if the location substantially affects the lot yield, consider alternative land plans. In most cases, however, the natural flow of the property can be used, and stormwater management facilities need not be detrimental to the project.

Stormwater management facilities can be used as a positive marketing tool. Detention ponds are valid open-space areas, and retention ponds can be designed with permanent water surfaces. (Premiums can be charged for lots bordering these facilities.) This approach could result in locating several stormwater management facilities within a subdivision and minimizing the number of inlets, pipe lengths, and stormwater management facilities.

Minimize pipe lengths by taking road drainage from inlets directly to natural drainage swales or stormwater management facilities. This eliminates multiple connections to inlets, which result in larger downstream pipe diameters. Larger pipe sizes increase the cost of construction.

Maximize the length of gutter flow and the spread of water in the street to minimize the number of drainage inlets. Many municipalities require the design of subdivision drainage systems to be based on a 25-year storm. This conservative design approach undermines the cost efficiency of residential subdivisions without a true benefit. Designing for a 10-year storm is adequate for housing developments.

Design a system using pipe sizes that minimize material and trenching costs. Pipe materials such as corrugated metal pipe could be an alternative, but this pipe is larger in diameter than reinforced concrete pipe (RCP) serving the same watershed. The cost in trenching and installation, combined with the problems associated with the availability of corrugated metal pipe, equals the cost of using RCP. Most municipal governments that accept the responsibility and maintenance of drainage systems in subdivisions prefer the use of RCP. Obtaining local government approval for alternative pipe material is seldom worth the effort, because it demands the use of additional corporate resources and delays the approval process, and results in only marginal cost savings.

The location of each stormwater management facility and detention or retention pond should benefit the maximum number of housing units in the subdivision. Ponds can be designed as entry features or as open-space features within a subdivision. In locations with high groundwater, a permanent water surface in a stormwater management facility can be maintained by using a shallow well to pump water into the pond. If over-excavation is also permitted, the pond area can be a source of fill material by the overexcavation.

Drainage facilities are the most regulated design elements of a housing project. The design engineer must devote time to this design aspect to reduce, minimize, and eliminate costly drainage facilities. However, the multiplicity of reviewing agencies has a direct effect on the engineer's ability to implement some of these ideas; it may take a devoted lobbying effort to change the subdivision regulations to include new proven designs.

Roadways

The roadway design can minimize earthwork by using the existing topography. If the roadway is in a cut, the housing units can be left above the roadway. This approach ensures that proper lot and street drainage can be accommodated cost effectively. Project costs can also be reduced if the site has material that can be used for road base and the material is permitted by the local municipality.

Perhaps the greatest cost savings can be derived from a reduction in pavement widths. Without question, the residential street-width requirements of most municipalities are conservative and overly restrictive for residential use. As a rule of thumb, the cost for a 30 ft-wide roadway and a 50 ft-wide right-of-way ranges from $150 to $200 per linear foot of improvement. Any reduction in pavement widths significantly reduces the per-linear-foot cost.

Most municipalities also have standard asphalt pavement requirements. Consider whether alternative pavement mixes could be less expensive, environmentally sensitive, and/or more readily available.

The type of curbing required for residential subdivisions is based on local use. Various design alternatives include asphalt rolled curb, standard L-block barrier curb, or concrete mountable curb. In the northeastern states, Belgian block curb is used in most subdivisions. The initial cost for Belgian block curb is higher, but the marketing appeal in a quality subdivision is tremendous. Also, the ease and cost of replacing damaged Belgian block curb are substantially less than for other curb types.

Communities in every state have debated the merits of public versus private roadways. Any opportunity to design single-family detached housing subdivisions with private streets should be considered during the project feasibility phase. This approach has wide market appeal, and the upfront cost of private roads is less than for public roads. On the downside, an HOA must be created to maintain and operate the roads, so a monthly HOA fee for the private facilities will be levied. Moreover, in single-family detached neighborhoods, the lot owners tend to prefer more independence, with fewer restrictions than HOAs usually impose and enforce.

Either way, the infrastructure must be of high quality to ensure that the level of maintenance is reduced. Quality infrastructure also reduces the cash reserves required of an HOA for replacement of infrastructure elements. If any portion of the infrastructure fails, the HOA must seek corrective measures from the contractor. To assure the quality of its work, the contractor should provide a two-year warranty on the infrastructure.

Quality Reviews

The design and construction options presented in this chapter are guidelines for creating a pattern of development that adds to the overall market appeal of the project while reducing costs. These ideas can only be implemented if the regulatory agencies agree with the design approach (Figure 11-2).

Once the design review is complete, the engineering firm produces final construction documents for submission to regulatory agencies. These plans include sufficient detail necessary for bid estimates to be obtained from land development contractors. The plans must comply with the regulations of the local municipal government. Otherwise, the plans can be deemed incomplete and returned with an extensive review list prepared by the reviewing engineer, thus delaying the project in the regulatory process. Any negotiations or discussions on design alternatives may or may not be resolved in the short term. Thus, review the design plans to ensure they are prepared to the best of the design group's ability. The project design may have a better chance of an approval on the

Figure 11-2. Design quality. Residential infrastructure design must balance the vested interests of the developer, government agency, and the home-buying public. A high-quality design approach will meet the needs of the stakeholders.
Source: David E. Johnson.

first submission if the regulatory agency's design preferences are used in the project.

Engineering Review

Appendix 11-1 helps the builder/developer in reviewing engineering plans before regulatory submission. This checklist is not intended to double-check the design characteristics of a plan—that is the responsibility of the registered engineer. Rather, this checklist can be used to verify that the information required by regulators is indeed in the plans. A concise review of the plans using this checklist ensures that the project's construction documents are as complete as possible before submission. The development goals should be reflected in the engineering plans and reports. As an example, the engineering plans should reflect the proper grading around the houses. The finished floor elevation relationship to the exterior grade must be consistent with a model building code.

Cost Estimate

After reviewing the engineering plans, the builder/developer should prepare an in-house land development cost budget for the project. Appendix 11-2 is a comprehensive land development budget form to be used at the preliminary engineering design phase. This budget document highlights key elements of the land development process. The preliminary estimate should be based on current unit prices and current material and labor prices. The comment portion of the form provides an opportunity to explain the cost approach taken for each line item. This estimate will be used in the contractual negotiations with land development contractors to verify their costs. It is also used in the project pro forma to verify the sales price for the lots or homes in the project.

By comparing the preliminary cost estimate versus the conceptual cost estimate (see Chapter 10), the developer/builder will derive significant insight on the project's ability to meet the market demands. The preliminary budget for land development also identifies items that could result in a cost savings. Areas that differ drastically from the conceptual land development budget are highlighted. This review and costing procedure allow the owner a chance to analyze the overall approach to the project and make changes for financial and market reasons. The opportunity to change the plans for financial or market reasons must be taken before the plans are submitted to the regulatory agency. Governmental inspection and escrow fees escalate during the review process of any subdivision plan. Thus, it is prudent business practice to carefully review the engineering plans before submission. Make sure the plans are consistent with company goals and identify any cost issues that may be drastically different than originally anticipated.

Appendix 11-1

ENGINEERING PLAN REVIEW CHECKLIST

COVER SHEET CHECKLIST

( ) Project title and location.
( ) Location map with north arrow and scale.
( ) Project area identified on location map.
( ) Key map.
( ) Legend.
( ) Engineers' notes, check for spelling and completeness.
( ) If notes are standard, make sure that the proper corrections have been made to make it applicable to project.
( ) Index of sheets.

GENERAL

( ) Check notes for completeness and spelling.
( ) Title blocks complete.
( ) North arrow and scale.
( ) Check with community subdivision regulations and engineering standards to make sure that all requirements are met.
( ) Proper cross reference is made between plan & detail sheets.
( ) Grading and drainage has been coordinated with adjacent properties.
( ) Adjacent properties identified.
( ) Check with record plat to make sure that all rights-of-way & easement are shown.
( ) Show block & lot numbers, also identify tracts, parks, waterways, street names.
( ) Show all property dimensions.
( ) Roadway and/or, driveway alignment is complete.
( ) Show bench marks if available.
( ) Show all existing information.
( ) Label existing information to remain, be removed, abandoned, etc.

PAVING & GRADING

( ) Show elevations at all inlets.
( ) Existing topo is shown, including contours at 1 ft intervals.
( ) Proposed grades are shown. Make sure that they are at or above flood criteria.
( ) Check that vertical curves are shown when streetgrade has a difference of over 1%.
( ) Show elevations at all low points, high points, break in grade, and intersections. Also show distance and slope between points.
( ) Roadways, driveways, bike paths, walkways are shown.
( ) Check that proper typical street section applies to road right-of-way.
( ) If applicable, check that proposed grade lines are properly shown.
( ) Check that finished floor elevation meet government requirements, building codes and yard drainage.

DRAINAGE

( ) Primary drainage corresponds with drainage master plan of the area.

( ) Direction of runoff is shown.
( ) All low points have provisions for drainage disposal.
( ) Label all pipes and show in pipe schedule (if one is used) the following:
    ( ) Pipe size.
    ( ) Pipe material.
    ( ) Length of pipe.
    ( ) Headwalls, if needed.
    ( ) Flow at peak if pipe schedule is used.
    ( ) Inverts, in and out.
( ) Check that pipe inverts and swale bottom elevations are coordinated.
( ) Check details for soakage pits, french drains, slab covered trenches, etc.
( ) Check for conflicts between drainage system and other utilities (water & sewer).
( ) Proper cover, provided on pipes.
( ) Identify type of inlets.
( ) Identify type of grates.

WATER DISTRIBUTION PLANS
( ) Check all notes for completeness and spelling.
( ) Title blocks complete.
( ) North arrow scale.
( ) Check with community subdivision regulations and engineering; standards to make sure that all requirements are met.
( ) Proper cross reference is made between plan and detail sheets.
( ) Show block and lot numbers, also identify tracts, parks, waterways, street names.
( ) Show all existing information.
    ( ) Existing water mains.
    ( ) Fire hydrants.
    ( ) Valves.
    ( ) Other utilities in area.
    ( ) Drainage structures.
    ( ) Buildings.
    ( ) Roadways.
( ) Check standard details to determine location of main with respect to r/w line.
( ) Label connections to existing water mains.
( ) If tapping, call out "wet tap and valve."
( ) Loop system if possible.
( ) Dead end lines should have blow-off valve at end.
( ) Show pipe size and material.
    ( ) Check regulations for minimum pipe size.
( ) Show valve locations.
( ) Show fire hydrants.
    ( ) Check regulations for maximum spacing.
    ( ) Minimum pipe size to fire hydrant to be six inches (6") diameter.
( ) Water service provided to all properties.
( ) Show all pipe fittings.
( ) Provide plug for pipes planned for future extensions.
( ) Check for conflicts.

- ( ) Vertical clearance.
- ( ) Horizontal clearance.

( ) Check for adequate cover over pipes.

SEWAGE COLLECTION SYSTEM

( ) Check all notes for completeness and spelling.
( ) Title blocks complete.
( ) North arrow & scale.
( ) Check with community subdivision regulations and engineering standards to make sure that all requirements are met.
( ) Proper cross reference is made between plan and detail sheets.
( ) Show block and lot numbers, also identify tracts, parks, waterways, street names.
( ) Show all existing information.

- ( ) Existing sewer lines.
- ( ) Other utilities in area.
- ( ) Drainage structures.
- ( ) Buildings.
- ( ) Roadways.

( ) Show pipe size and material.
( ) Show length (center to center between manholes) and slope of sewer line.
( ) Show trench cuts.
( ) Show manhole and manhole number.
( ) Indicate top of manhole elevation.
( ) Indicate invert elevations of line at manhole.
( ) Show direction of flow.
( ) Locate (by stationing) all sewer laterals.
( ) Show location of sewer line with respect to right-of-way line.
( ) Show location of lift stations.
( ) Show location of treatment plant if possible.
( ) Show on plan view details of sewer and drainage system crossings.
( ) Indicate the minimum allowed cover.

DETAIL SHEET

( ) Complete title block.
( ) Check all standard details and complete by adding any particular information which may be needed.
( ) Detail titles and identifications are correct.
( ) Proper cross reference is provided between plan and detail sheet.
( ) Check general notes for completeness and spelling.
( ) Typical roadway and or driveway sections if different from community engineering standard.
( ) Pavement design.

- ( ) Surface course, type and thickness.
- ( ) Base course, type and thickness.
- ( ) Subbase stabilization.

( ) Pavement slopes.
( ) Pavement width.
( ) Sidewalks.
( ) Curbs and gutters.
( ) Swales.
( ) Property lines.
( ) All features properly dimensioned.
( ) Typical grading details if used.
    ( ) Lot grading.
    ( ) Cul-de-sac grading.
    ( ) Green areas.
    ( ) Block grading.
( ) Show section for waterway, lake, ditch, retention basin, etc.
    ( ) Maximum side slopes.
    ( ) Grading up to the property line.
    ( ) Rights-of-way.
( ) Detail all drainage structures and systems which are not part of community engineering standards.
    ( ) Inlets.
    ( ) Grates.
    ( ) French drains.
    ( ) Soakage pit.
    ( ) Slab covered trench.
    ( ) Control structures (weirs, culverts, etc.).
( ) Check headwall size required and make sure that the area around the structure has been graded properly.
( ) Provide details of erosion protection systems if required.
( ) Protective slab detail for pipes with less than minimum cover.

WATER SYSTEM
( ) Thrust block details.
( ) Typical meter-service assembly.
( ) Blow off valve detail.
( ) Gate value and box detail.
( ) Wet tap and valve detail.
( ) Sampling point details.
( ) Typical pipe, valve and hydrant location detail.
( ) Typical hydrant and valve detail.
( ) Special crossing details.
( ) Protective slab and anchoring for pipes with less than minimum cover.
( ) Detail water and sewer crossing if less than minimum clearance is provided (Vertical. 18 in. and/or Horizontal: 10 ft).
( ) Trench backfill detail.
( ) Pavement restoration detail.

SEWAGE SYSTEM

( ) Standard lift station detail sheet.
- ( ) Location map.
- ( ) Typical wet well installation.
- ( ) Wet well sections.
- ( ) Wet well plan view.
- ( ) Wet well notes.
- ( ) Pump data.
- ( ) Electrical notes.
- ( ) Electrical diagram.
- ( ) Force main data.
- ( ) Bill of materials.

( ) Brick manhole.
( ) Precast manhole.
( ) Manhole frame and cover.
( ) Standard drop connections.
( ) Sewer riser detail.
( ) Sewer lateral detail.
( ) Terminal clean out detail.
( ) Protective slab detail for pipes with less than minimum cover.
( ) Trench width details.
( ) Trench backfill details.
( ) Bedding details.
( ) Pavement details.
( ) Air release valve.
( ) Detail water and sewer crossing if less than minimum clearance is provided (vertical 18 inches and or horizontal 10-ft).
( ) Show fire hydrants.
- ( ) Check regulations for maximum spacing.
- ( ) Minimum pipe size to fire hydrant to be six inches (6") diameter.

( ) Water service provided to all properties.
( ) Show all pipe fittings.
( ) Provide plug for pipes planned for future extensions.
( ) Check for conflicts.
- ( ) Vertical clearance.
- ( ) Horizontal clearance.

( ) Check for adequate cover on pipes.

1 in. = 25 mm; 1 ft = 30 cm

Appendix 11-2

Pro Forma Land Development Budget

			# of Units	
PROJECT :				
ITEM	PRELIM	REVISED	BID	BUDGET
	DATE	DATE	DATE	DATE
SURVEYING				
Boundary & Topo Survey				
Roadways & Sitework				
Curbs & Gutters				
Building Pad Layout				
Sanitary Sewer Layout				
Force Main Layout				
Water Main Layout				
Drainage Layout				
Sidewalk Layout				
As Builts				
Clearing				
Restaking				
Other Surveying				
TOTAL SURVEYING				
ENGINEERING				
Design On-site & Off-site				
Plat Preparation				
Record Plat				
Testing				
Inspection & Certifications				
Drafting				
Printing Charges				
Quantity Takeoffs				
Other Engineering				
TOTAL ENGINEERING				

PROJECT	PRELIM	REVISED	BID	BUDGET
	DATE	DATE	DATE	DATE
DEVELOPMENT FEES				
Permits & Submission Fees				
School Fees				
Park Dedication				
Other Impact Fees				
Electric Fees				
Water Connection Fees				
Sewer Connection Fees				
Cable TV				
Telephone				
Inspection Fees				
Construction Bonds				
Escrows				
Other Fees				
TOTAL DEVELOPMENT FEES				
MOBILIZATION				
Trailer Delivery & Setup				
Electric Deposit & Hookup				
Telephone Deposit & Hookup				
Water & Sewer				
Temporary Road & Parking Area				
Barricades				
Landscaping				
Interior Improvements				
Exterior Improvements				
Signage & Barricades				
Permits				
TOTAL MOBILIZATION				

PROJECT	PRELIM	REVISED	BID	BUDGET
	DATE	DATE	DATE	DATE
SITEWORK				
Clearing & Grubbing				
Grassing & Mulching				
Tree Removal				
Trash Removal				
Rock Removal				
Lake Excavation				
Strip Topsoil				
Rough Grade (Cut & Fill Balance)				
Finish Building Pads				
Hauling Material Off-site				
Import Fill Material				
Demucking				
Testing				
Other Sitework				
TOTAL SITEWORK				
STORM DRAINAGE				
Pipework				
Headwalls, Flared End Sections				
Inlets, Manholes				
Outlet Structures				
Rip Rap				
Other				
TOTAL STORM DRAINAGE				

PROJECT :	PRELIM	REVISED	BID	BUDGET
	DATE	DATE	DATE	DATE
SANITARY SEWER				
Pipework				
Manholes				
Force Main				
Lift Station				
Service Connections				
Connection to Existing Line				
Testing				
Other				
TOTAL SANITARY SEWER				
WATER DISTRIBUTION				
Pipework				
Valves, T's, Crosses, etc.				
Fire Hydrants				
House Connections				
Connect to Existing Line				
Testing				
Other				
TOTAL WATER DISTRIBUTION				
TEMPORARY UTILITIES				
Electric				
Water				
Sewer				
Other				
TOTAL TEMPORARY UTILITIES				

PROJECT :	PRELIM	REVISED	BID	BUDGET
	DATE	DATE	DATE	DATE
PAVING				
Cut Roads				
Grade, Spread & Compact				
Stabilized Subgrade				
Asphalt Base				
Asphalt Surface Course				
Resurfacing				
Leveling Course				
Milling Operations				
Signs & Stripping				
Curb & Gutter				
Stone				
Other				
TOTAL PAVING				
LOT PREPARATION				
Lot Clearing				
Remove Unsuitable Material				
Deliver Fill to Lot				
Cut & Fill Operations				
Pad Construction				
Flatwork Grading				
Final Grade				
Driveway				
Driveway Apron				
Repair Road at Connection				
Other				
TOTAL LOT PREPARATION				

PROJECT :	PRELIM	REVISED	BID	BUDGET
	DATE	DATE	DATE	DATE
R/W & COMMON AREAS				
Street Trees				
Entry Feature Landscaping				
Entry Features, Walls, Fences				
Sidewalks				
Irrigation System				
Other				
TOTAL R/W & COMMON AREAS				
FINAL SURVEYING				
Plot Plan				
Clearing Limits				
Grade Stakes				
Building Stakeout				
Foundation Location Plan				
Final Survey				
Monuments				
Other				
TOTAL FINAL SURVEYING				
SUBTOTAL DEVELOPMENT COST				
LAND DEVELOPMENT CONTINGENCY				
OVERHEAD				
PROPERTY TAXES				
INTEREST				
MISCELLANEOUS				
TOTAL COST				
TOTAL COST PER UNIT				

Chapter 12

Residential Land Development Standards

This chapter analyzes the variety of residential land development standards used in different areas of the United States and presents a set of criteria that can be used by practicing engineers. The data on land development standards in this chapter has been provided over the course of many years by participants in a seminar on residential land development practices, sponsored by ASCE. The participants include design engineers, regulatory engineers, and development engineers, and the criteria apply to roadways, water and sewer lines, and drainage.

The criteria that engineers must use to design residential streets are inconsistent, and some of the standards have no engineering basis. Many standards are developed locally and modified politically to establish local control over the residential land development industry. Some states require that regional residential street standards be used by all municipalities within that region. Engineers and regulators must evaluate a locale's current standards against the following criteria to ensure that the local standards are efficient and cost-effective for protection of the health, safety, and welfare of the citizens.

Although certain health, safety, and welfare issues must be considered in developing appropriate engineering standards for residential subdivisions, these issues must be balanced against the cost–benefit to the initial user and to the ultimately responsible party—the local government. The cost of housing increases as standards become too conservative, and the cost to develop residential subdivisions continues to rise.

Residential Streets

The street standards imposed by most municipalities are still oriented to vehicles rather than pedestrians. Historically speaking, the development of residential streets has resulted in a "bigger is better" philosophy, which is a dramatic departure from the way our cities were originally planned and developed. Primarily, the standards we use today are a result of the use of cars and our standard and style of living. This suggests that the engineering criteria used to design streets should be somewhat similar across the country, although with obvious localized changes. A consensus derived from the compilation of design data and discussions with engineers across the country establishes the following residential street design standards as the basis on which to prepare an ordinance for land development design.

Right-Of-Way

Most municipalities use a 50 ft right-of-way for local residential streets. The right-of-way changes as the number of lots or average daily traffic (ADT) changes for a road segment. However, several municipalities require a 60 ft right-of-way for a local subdivision street. From a design point of view, the additional right-of-way does not significantly protect the public welfare and results in the removal of additional land from the tax roles.

Pavement Width

The participants of the survey indicated more than 20 different residential street widths. The most common standard used for pavement width of a local residential street was 28 ft, which can be substantiated by having an 8-ft parking lane on one side of the street and two 10-ft travel lanes.

It is important for a municipality to establish a hierarchy of roadway widths to accommodate vehicular movement, in which case the 28 ft width could be the standard for a residential subdivision. The width would increase and decrease as the number of units using the road varies. The hierarchy of road widths should be based on ADT values for residential use. This ADT value should be established at the local level, but it would not be unreasonable to have a threshold of 1,500 ADT for the 28 ft pavement width.

Cul-De-Sacs

The most common dimensions for a cul-de-sac are a 50 ft radius right-of-way and a 40 ft pavement radius. The length of a cul-de-sac varies considerably, but the consensus is between 600 and 1,200 ft. The reason for the variance is that the length of a cul-de-sac is set by local preferences. Many regulations impose a maximum number of units to be served by a cul-de-sac; the most common number is 25 homes.

Curbs and Sidewalks

Most new residential streets are required to have curbs, which provide for pavement protection and drainage conveyance. These streets also have 4 ft sidewalks on both sides of the street.

Vertical and Horizontal Curves

A local ordinance should provide the minimum and maximum grades for vertical curves. Most jurisdictions require 0.5% minimum and 10% maximum grades for residential streets. Certainly, there are many states in which the 0.5% grade may be too steep, and the 10% grade may not be steep enough. Depending on the topography of the area, these standards should be modified to ensure that the engineering design can easily accommodate the existing lay of the land, minimize earthwork, and potentially save more of the existing vegetation on site. The maximum grade within 100 ft of a street intersection should be 6%.

The minimum horizontal curve for residential streets should be 100 ft. This criterion provides engineers with design flexibility and does not jeopardize the health, safety, and welfare of the users. The horizontal alignment dictates the speed of the drivers in residential applications. The goal of residential street design is to put more emphasis on the pedestrians and homeowners and less on vehicle movement. Residential street design should not be based on a design speed; it should be designed to limit vehicle speed within a housing subdivision. The most common pavement cross-slope is 2%.

The angle of an intersection is not mandated to be set at 90 degrees. Most ordinances allow an intersection of not less than 75 degrees. Intersection design for residential uses should permit this flexibility, but designing an intersection at less than 90 degrees should be the exception and not the rule.

On major highways and collector roadways, tangents between reverse curves are important in the design of the horizontal alignment. However, in the design of residential streets where slower speeds are encouraged, a tangent between reverse curves becomes less important.

Water and Sewer Lines

Throughout the United States, it is common practice to set a minimum pipe size of 6 and 8 in. for both water and sewer lines. In most jurisdictions, PVC pipe is the preferred pipe material for both water and sewer lines. The distance between fire hydrants is 500 ft. The distance between sewer manholes is 400 ft.

Drainage

As an element of residential infrastructure design, drainage remains very site-specific because of the climate differences across the country. Also, there seem to be many overlapping reviewing agencies that at times have different criteria for designing drainage facilities for residential subdivisions. Most ordinances require stormwater management facilities to be addressed on a project basis; this results in the maintenance of isolated systems adding to the overall cost of community drainage facilities. Most communities have not instituted regional stormwater management planning. Jurisdictions that have planned to control stormwater on a watershed basis have found benefits in centralizing the facilities for maintenance purposes.

The intent of stormwater management is to control the runoff generated by a subdivision after development and to discharge it at a rate less than or equal to the rate generated before development. Roadway drainage facilities are designed using a 10-year storm event. The maximum spacing between inlets is 400 ft. Some jurisdictions use the spread of water in the travel lane to determine the inlet spacing. The minimum size for a storm drainpipe in most jurisdictions is 15 in.; however, some communities permit a 12 in. pipe, and others require a minimum pipe size of 18 in.

Chapter 13

Project Approval Strategy

All housing projects have an impact on the community in which the subdivision is located. In most cases, the impact on the immediate area is positive. Therefore, during the regulatory process, the builder/developer should initiate a proactive public relations program to involve, inform, and educate local residents about the benefits of the proposed project. Engineers designing a new project have a responsibility to develop plans that are technically sound but also cost effective for the builder/developer.

Being sensitive to existing neighborhood concerns helps to minimize local opposition to a project. The builder/developer must investigate and seek to understand the needs and priorities of the community. By doing this, the project can be presented as a "good neighbor" during public hearings, the regulatory permitting process, the sale of homes, and after homes have closed in the subdivision.

The builder/developer must maintain a commitment to communication that will convey the positive aspects of the project to the public. Communication enhances the company's ability to gain support from the community, and this general positive awareness is beneficial in obtaining government approvals. Neighborhood support also allows the homes to be introduced into the marketplace in a timely manner. At the initial sales release, the commitment to communication creates a positive environment to attract home buyers.

Approval Strategy for Builder/Developers

A successful strategy for gaining project approval from both the community and government regulators is derived from careful preparation in several areas, which are considered next. Many builder/developers take these areas for granted, missing an opportunity to proactively manage issues that can ultimately determine the success or failure of the project presentation before planning boards.

Experts

Design professionals and consultants often are highly proficient at the design and technical aspects of a housing project, but seldom do they possess the ability to convey that message before a public audience. In getting to know the consultants, evaluate their ability to take technical information, dissect it, and present it in lay terms. Your experts must not talk down to the general audience and planning board members or overwhelm them with technical jargon.

If the project design is technically acceptable but cannot be adequately conveyed, planning board members may impose unacceptable conditions or vote against the project. Thus, observe your consultant's presentation of a project during a public forum. At the very least, meet with your consultants for an extended time so you can assess the way they express their thoughts and convey technical information to audiences.

Political Problems

Housing projects require strategies for gaining government acceptance of the subdivision. Examine the political issues of the community if you anticipate that local government wishes to have its problems resolved by new developments. By understanding local government's problems, you can incorporate solutions into the design and convey answers to the problems during the public hearing. As an example, if the volunteer fire departments, emergency services, or first aid groups do not have a sufficient number of volunteers or equipment for full service, the approval of additional housing projects in the area would magnify the problem.

The tax consequences of changing from volunteer to paid services pose a political problem for the local planning/zoning boards. If your project is caught in the middle of this, government's inability to resolve the issue may result in failure for your project. Know the issues and have possible answers to the problem before your board presentation so you can resolve the political issues and possibly avert delays.

The housing industry is complex because it is regulated at the grassroots level by local politics and local government agencies. The technical compliance for single-family subdivision design is usually clear cut. Differences in engineering approaches may be the only debate requiring a compromise in the approval process. The local, regional, state, and federal political arenas, however, influence the design characteristics of a subdivision beyond the technical parameters guiding infrastructure design. The best way to understand the political atmosphere or environment is to get to know the regulatory staff and the power base of the planning board. Staff members will be able to outline the pros and cons of a proposed housing development. They can also define the political agenda and explain what the community is requiring of new housing projects during the approval process.

Community Relations

A community relations program that builds support from the public can maximize the support from elected officials and business leaders for the project (Figure 13-1).

Figure 13-1. Community involvement. Developers of new housing projects must seek involvement from adjacent neighbors and community advocates. The consensus approval process is designed to address public concerns.
Source: David E. Johnson.

This type of program can be accomplished by implementing the following strategies:

- Direct communications, such as letters, community meetings, flyers, and introductory material, to inform the community as a whole;
- Low-key neighbor-to-neighbor public relations programs to seek feedback from residents bordering the project so they become part of the design and not part of any problem;
- Community citizen engagement in open forum stakeholder meetings, public sessions, individual interviews, and completion of online surveys; and
- Social media platform posts to inform citizens and seek feedback.

The neighbor-to-neighbor strategy has the advantage of being conducive to an excellent exchange of ideas. It is also effective in nurturing a sense of cooperation. However, use caution: if the public relations campaign becomes excessive, the audience maybe become more skeptical of the project itself.

Keep an open line of communications with community leaders and elected officials. Designate a staff member or the owner of the company as the contact person to elevate the importance of the project in the eyes of the community. Any presentation before the community should be concise, on track, and meaningful in substance. All public campaigns must address the town's concerns and issues so that a sense of positive momentum can be achieved before the planning board hearing.

Strategic distribution of press releases to the local media, whether print or television, is an appropriate way to disseminate information to dispel any community rumors about the project. Do this only after, and to complement, scheduled meetings with neighbors, elected officials, and community leaders.

If the property and project become controversial, create an advisory group with town leaders as members. The chair might be a local business leader, mayor, former town council member, or a civic leader. This advisory committee would be charged with the goal of disseminating project information in a proper forum, and constituents would listen to their points of view during regulatory hearings.

As with any business, the builder/developer should establish a presence in the community by being involved in the events of the community. Donating time, effort, and resources to the betterment of the community enhances the reputation of the firm and provides positive momentum. Community relations programs succeed by striking a balance between art and science. The more positive attention the community receives, as a whole, the more likely assistance will be given in obtaining an approval of the project.

Project Costs

New housing projects have been and will continue to be a conduit for improving or replacing outdated community-owned infrastructure. Local communities cannot always afford to upgrade existing infrastructure to accommodate housing growth. Thus, communities reequest new housing projects to assist in a fair-share approach for improving infrastructure in and around the new housing project. The mechanism to gain this development support is through the planning board process. Many times, these issues are not included in discussions with staff members but are a result of the political review of the project. When you create your project pro forma, be sure that all project costs are well defined, contingencies established, soft costs budgeted, and profit margin objectives maintained. Politics will affect your housing project, and you must know your costs and know what you can afford and are willing to let go.

Project presentations before planning boards are the forum in which local officials ask developers to provide construction or financial assistance toward the upgrading of existing infrastructure. At that moment, in front of the planning board, you must decide whether to provide assistance or to debate the issue and seek relief from their demands. If the project costs are known, you can make your decision based on the project's financial facts. Know the cost of off-site improvements, that is, road widening, drainage facilities, streetlights, and so on, so you can render a decision before the planning board. Weigh the legal costs and delays versus the costs of site improvements. There are no winners in legal confrontations at the planning board table. If the project can support the planning board's requests, a compromise is the best solution.

Planning Board Presentations

Project approval is contingent on the ability of the builder/developer or design team consultants to answer questions under the scrutiny of each planning board member, neighborhood group, and the public. Prepare outlines for each presentation and orchestrate the design professionals to present the project in a logical order and concise manner.

Answering all questions, asked and not asked, of the design team during the presentation makes a positive impression on the planning board. Try to resolve controversial issues outside the public hearing. Minor issues can be resolved during the presentation and entered into the record.

Whenever an issue is discussed in the public forum and it cannot be satisfactorily resolved by the design team, the individuals and groups voicing objections benefit. In addition, the planning board members are faced with a political dilemma, the staff members put in a questionable position, and the design professionals are in a compromising situation. Most likely, the approval will be delayed until all outstanding issues are resolved.

A planning board presentation strategy is successful when the project obtains approvals and construction can commence without delays. In most states, a delay at a public meeting because of unresolved issues could effectively delay the project by one month. A one-month delay may change the project schedule by more than one month. Any political delay could put your project in jeopardy if circumstances—such as site conditions, adverse weather conditions, and so on—affect the start of the project after the issues have been resolved.

Role Playing

Two weeks before the planning board, board of adjustment, or council public meetings, schedule a meeting with the design team to present the project specifics before a select audience. Each presentation should be specific, to the point, and easy to understand. At this meeting, select the members who present the project parameters in a style that does not sound adversarial or condescending. During each presentation, the builder/developer should act as the mayor and seriously guide the role-playing scenario. Role playing should be as real as possible. The "mayor" should question all issues. Challenge the experts to think on their feet and to concisely answer all questions asked. Have several other staff members attend and ask specific controversial questions to all presenters.

This format energizes the design professionals into a mental state of readiness, and it also achieves nonbiased discussions on any political issue that may be lingering. All issues that could have an adverse effect on the project approval should be discussed at length. During this role playing, the design team should use visual aids. Judge their approach, style, and content against the goals established for the project.

Encourage all design team members to answer questions completely, but without offering more information than is required to answer the question. Know what your design team will say, when they will say it, and why they will say it. Conclude the presentation with corporate positions on issues that could be raised at the planning board meeting. All issues resolved during the presentation will help achieve a successful affirmative vote.

Planning Board Meeting

Schedule the design team to participate in a dinner meeting two to three hours before the planning board meeting. This permits everyone to focus on the task at hand for that evening. Discuss any last-minute strategies or issues that might surface during the presentation and public session. All handouts and exhibits must be available at the dinner meeting. The meeting's format provides a relaxed environment for last-minute debate of issues, which should be openly expressed and discussed at length. However, once the decision is made on how to handle each specific issue, the design team must adhere to the consensus of the group. A united professional team that believes in the project provides positive emotion that translates into momentum for the project.

Engineers and Public Hearings

As a key member of the project design team, the engineer has an especially important role in supporting a residential housing project as it moves through the regulatory approval process. This section presents 10 rules to help an engineer contribute to a builder/developer's success in the public approval phase of a project.

Rule 1: Know the Political Hot Buttons

Local politics control the development process. Every community has several key political "hot buttons" that govern the actions of the elected officials. The hot buttons could be no-growth policies, overcrowded schools, lack of volunteer firefighters, lack of affordable housing, or environmental issues, among others. These hot buttons may change, but the old issues never fade away. The obvious hot buttons must be addressed in your project plans. The hot buttons you must know are the most current issues affecting your community at the time your project is being considered by the planning board.

The successful engineer will answer all questions raised at the planning board and obtain a positive decision at the very first meeting. If an unforeseen hot button issue arises that has not been contemplated by the builder/developer and was not resolved before the meeting, the project will be delayed at least one month or perhaps longer.

How Do You Find Out About the Issues?

Political involvement at the local level may be necessary. Join the chamber of commerce or become politically active. Stay involved at the local level by talking with staff, board members, and politically active residents. The staff members assigned to the project are responsible for implementing the hot button issues and bringing those issues to the forefront of discussions with the applicant. Become professionally attuned to the requirements of the staff. Understand the political emphasis by local government at the time your project is being reviewed. Without question, community hot buttons have a cost implication to the project and perhaps may delay it. Address the hot button issues before the issues are raised in a public forum.

Should You Meet with Board Members?

In a perfect world, reviewing the project with board members ahead of time would be an excellent way to inform them of the project specifics and help them better understand the approach to the project and the design emphasis, benefits, and positive implications of the project for the community. However, in many locales, meeting with board members is not appropriate because the public would question the ethics, especially in a highly charged political arena. Also, sunshine laws—those that require transparency, as well as disclosure, in government and business—may restrict council members from meeting outside of open public sessions. The project plan should rest on its own merits. "Selling" the plan to council members and town staff would not be necessary when engineers, developers, and/or builders understand that reputation matters.

The best regulatory process at the local level is the use of a site plan review committee or a development review committee, which permits staff members and selected planning board members to meet on a regular basis with builder/developers and their engineers on upcoming projects. It is an informal hearing in which the project can be discussed in a casual atmosphere, and the board members and staff members can voice their opinions. Most projects can be modified to meet the concerns raised at the informal hearing. Gaining immediate feedback on the project design and project issues is important so that all issues can be addressed before the public hearing.

Should the Engineer Become Politically Involved?

A revenue source for engineering firms can come from being a municipal engineer, designing public works projects, or being retained to review projects by private developers. The municipal services provided by an engineering firm are somewhat recession-proof and are an excellent addition to any business plan. In some locations, to obtain this position, engineers need to become politically involved. Otherwise, engineers can stay at arm's length from the political arena and allow the technical design to be judged on its own merits. The best practice is to keep the engineering aspects of the land development business technical in nature and without political influence, if possible.

Rule 2: Public Participation Is a Key Component

Dealing with the public at planning board meetings has become much more difficult and will only become more so as growth patterns increase. In the building business, profitability is directly influenced by the public. The approval system allows public input and public scrutiny for every project. Planning board members, all too often, assume the public views are correct and the development industry must prove the design approach is proper. It has become increasingly important, regardless of the size of the project, for engineers and their clients to reach out to neighborhoods and adjacent property owners to present the project prior to a board meeting.

Gaining neighborhood input before a planning board hearing takes place is advantageous, because it provides insight on the neighborhood's problems and concerns with the project. Project supporters may also be discovered. In addition, neighborhood input provides the opportunity to address public concerns before the public debate. Most likely, the neighborhood views and the views of the builder/developer are not going to be consistent. Therefore, it is better if the builder/developer can stand before the planning board and say that meetings were held with neighbors to explain the project, thus publicly declaring that community views have been taken into consideration. Most public hearings are not an appropriate environment for explaining the details of a plan. Sometimes, the reaction in a public forum will be negative because a project without concessions from the developer is not acceptable.

If a project is of significant size, it will cause a physical effect on the surrounding community and can strain community services, such as the fire department, emergency response, police, schools, and so on. In this case, a positive public relations approach may also be necessary. This public relations program would consist of positive advertisements and media outreach to "spin" the positive aspects of the project. It should also illustrate what positive ramifications have on the community as a whole and include public participation in small neighborhood meetings.

Always consider the local neighbors and adjacent property owners' views and gain their input before a public planning board hearing. Your project will be viewed more favorably by the planning board members if your outreach program is sincere. Also, impacts of negative public input can be defused by addressing those public concerns during the presentation.

Rule 3: Know Your Costs

If you do not know the costs of your project before a hearing, you will not know what could be cut or where to compromise when asked by the planning board members. Knowing the cost of development means you can make sound business and political judgments at the moment when planning board members are seeking concessions or compromises on the project. Any decision to comply with a planning board request will affect the cost of the project. If you do not know your costs and respond by saying, "I have to check" or "We'll get back to you on that," you will be delayed at least one month, and time is money.

Every hot button in a community has cost ramifications. Knowing the hot buttons and the costs associated with the hot buttons allows the board and builder/developer to reach an agreement in a public forum after constructive debate and concessions at the time of the meeting. If you cannot address the issues before the meeting, resolve hot buttons or issues of public concern during the meeting; that way, your project will remain on schedule, and unusual or undue delays will be avoided.

Engineers should provide a detailed cost estimate of the project infrastructure at the preliminary plan stage. Then the builder/developer can analyze the cash flow requirements and make an assessment of the project viability before seeking final approvals. Regulatory standards governing the engineering design are not flexible and changing the engineering design after the preliminary design is complete will not result in significant infrastructure cost reductions. Most municipalities are not receptive to modifying their design standards. If design variances are required, seek regulatory approvals before the preliminary design plans have been completed. Engineers affect the costs of a subdivision project on a daily basis and directly influence the profitability of the company. Adding a buildable lot, reducing infrastructure costs, or expediting the plan production or regulatory approval process all have a positive financial effect on the project.

Rule 4: Prepare for a Board Hearing

To prepare properly for a planning board hearing, you should know what the community hot buttons are, talk to neighbors, and understand the project costs (Rules 1, 2, and 3). You should also become familiar with how each community's planning board staff operates and how its board functions to avoid surprises on the day of the meeting.

Engineers presenting a project before a planning board should find out in advance

- How the hearing room is laid out;
- Where the board members, audience, and presenter are situated in the room;
- How the exhibits will be situated for presentation;
- What the acoustics of the room are like and whether a microphone is used; and
- Whether an easel or a computer is available to display graphics.

When presenting, always repeat the questions you are asked so that the entire audience can hear both the questions and the answers. Every jurisdiction is different in the way it processes plans, reviews plans, and debates the designs. Engineers must understand the nuances of each community—builder/developers will retain the engineers who understand their communities the best. Engineers should establish themselves as most knowledgeable about a jurisdiction and develop strong connections to the staff and politicians, so the builder/developers have no better choice but to hire them.

Rule 5: Prepare Quality Graphics

Today, computer-generated graphics for presentations at planning board hearings are widely available, yet many presenters fail to use exceptional color graphics to illustrate the characteristics of their projects. Certainly, the more sophisticated or advanced the planning board, the easier it is to display the specifics of the project through computer presentations. Appropriately colored exhibits are just as important as the presentation itself. The illustration of a project elevates the professionalism of the design team members and earns a higher level of attention from the planning board members and the public. Learn effective styles of illustration yourself or have a staff member with artistic talent prepare the presentation material.

Rule 6: Use Appropriate Tools of the Trade

To prepare for a planning board hearing, the design team members should practice their presentations by role playing. The project may be of such importance or controversy that written scripts may be necessary. The written scripts should include anticipated questions and prepared answers. Certainly, do not allow the presentation to be spontaneous. Make sure the message is well-thought-out and well-presented. Role playing allows the design team members an opportunity to critique the presentation and ask difficult questions about the project. Practice answering difficult questions in a positive and upbeat manner. Try not to use notes during the presentation but have project material at your fingertips, if needed. Always arrive at the hearing ahead of time and review the agenda to determine if your position on the agenda is favorable. Repositioning the project on the agenda can be accomplished without too much difficulty. Having the project at the end of the agenda may be beneficial because the board members become tired. Also, have the project follow a controversial project, so those in attendance unleash their displeasure on other projects. In addition, most protestors will leave after the controversial project has been dealt with by the board.

Successful professional consultants can explain a technical subject in lay terms. Most planning board members and the public are not trained to read construction drawings or discuss technical issues, so they primarily address emotional and political issues. All engineering firms must train their young engineers in the art of presenting a technical project before a planning board in an environment that could be hostile, adversarial, and stressful. Builder/developers should be willing to mentor and encourage young engineers in representing them before a public body. Most clients will allow an engineering firm to give a young engineer the opportunity to present, although some clients prefer to use only the experienced presenter to obtain approvals.

Rule 7: Orchestrate the Performance

As the number of housing units in the project increases, orchestrating the design team's planning board performance becomes more important. First, determine the order of presentation as well as the role of the client and the attorney. Many builder/developers present their own projects and call on their engineers only for technical support. In most jurisdictions, however, the engineer is the primary presenter of a subdivision project, and the client is there in a support role and to participate in making decisions.

The design team members should arrive before the meeting starts with all graphics. Keep each presentation as brief as possible. The last presenter should be the presenter of the "sizzle." The sizzle focuses on a subject matter that everyone can relate to and can feel good about (for instance, landscaping, architecture, colors, environmental sensitivity, amenities). The sizzle can sell the project to the board members.

Rule 8: Present a Project with Enthusiasm

It is helpful to show enthusiasm for the project design. In fact, most engineers are so intimately and emotionally involved in the design of a subdivision that they can easily take exception to criticism and critique of the project. Keeping discussions professional and impersonal will best serve the engineer at the planning board podium. Display enthusiasm and humanize the project. This softens the delivery of the technical characteristics of the project and is better received by the public and board members. It also helps to establish your credibility and lead the audience to view the project more positively. Be genuine and sincere in the presentation and the message.

On very few occasions, a presenter must show anger during a presentation or during a question-and-answer session. These rare occasions are precipitated by a project being judged politically rather than on its technical merits. But 99.9% of the time, showing anger at a planning board hearing is not warranted, and if you do so, your credibility will be lost. Neither the project nor the presenter benefits from a display of temper in a public forum. Do not show distress or nervousness in your voice, and certainly do not display arrogance. Local approvals allow input from all people, so regard all input with a high degree of professionalism. Success at planning board meetings hinges on the ability to present the project in a distinctly professional manner founded on enthusiasm.

Rule 9: Answer Questions Briefly

Engineers should consider a planning board hearing as a court of law. Thus, tailor your presentation and answers as if you were giving a deposition. If a question demands a yes or no answer, then answer yes or no and do not elaborate or explain. Elaborating gives the audience and planning board members an opportunity to think of more questions. Always repeat the question you are asked, so that everyone hears it and that you are sure you have interpreted the question correctly. The most important period during the hearing is when the public is permitted to ask questions. Answer these questions respectfully and show sincere interest in the public's comments and questions. Answering questions in a way that makes the public feel confident about your answer is an art that comes only with practice, practice, practice. Because many planning board members react to the response of the public, a positive framework in which to answer questions must be maintained by everyone on the design team.

A negative and adversarial audience causes the project to become more controversial, and the planning board members are less likely to embrace the project for its positive contributions to the community. The presenter must defuse an adverse public attitude before the meeting, during the presentation, and during the question-and-answer session.

Rule 10: Pass the Torch

Engineering firms should implement an in-house program for training young engineers in the art of presentations to planning boards. Most engineering students newly graduated from college are not trained in giving speeches, especially not before an adversarial planning board. Engineering firms should consider the following training program:

- Have the young professionals take a communication class, such as Toastmasters or Dale Carnegie.
- Develop an in-house speaking forum for the young engineers to present a project to other staff members in a role-playing atmosphere. The role playing should include an intensive question-and-answer period.
- Have the young professionals attend planning board meetings to get a feel for the action and experience the way meetings are organized. Have the young engineer attend a planning board hearing on a controversial project and attend every meeting on that project until a vote is rendered.
- Have the young professional present a small noncontroversial project before a planning board.
- Train the young professional to communicate easily with people in planning board presentations and public participation programs by having them present a technical subject in lay terms.

A methodical training program allows young engineers to progress more rapidly and better serve the client base in and out of the planning board hearing room.

Engineers at the Crossroads

The engineering community associated with the housing industry is certainly at a crossroads. In one direction is the challenge of providing infrastructure design that protects the health, safety, and welfare of future home buyers. In the other direction is the challenge to ensure that each design element is cost effective. When government agencies insist on standards that do not justify the benefit to society, the engineering community must counteract overregulation.

When governmental engineers require infrastructure that is conservatively designed, the cost of developing land increases, and this cost is passed on to the consumer. Higher-priced homes and homeowner responsibilities for infrastructure maintenance are a direct result of a local government that dictates excessive standards without fully understanding the consequences. Many excessive regulations add to the constant monthly cost of operating a home. As an example, regulations may mandate a privately owned stormwater management facility in a single-family subdivision, which means, in turn, that CCRs will be needed to govern the facility. The costs of maintaining the facility and enforcing regulations will be added to the costs of purchasing a home. In effect, local municipal governments are sidestepping community growth issues and passing the costs on to new home buyers.

The engineering community can directly affect the way housing developments are designed in the future, especially when regulatory efforts continue to correct residential infrastructure problems with ever more conservative design approaches. All residential design engineers should use design standards that are practical and cost effective. Use standards that justify the housing project intent, regardless of the consistency with written governmental standards. Most municipal design standards are minimum–maximum regulations. Designers should not be hindered or confined by inflexible design standards. The land form and housing market should be sufficient for granting design variances.

Roadways seem to be designed wider, drainage pipes seem to expand in size, and design storm events seem conservative (i.e., design of the storm drain system is based on a 25-year storm rather than the standard 10-year storm). Public and private engineering firms must reach a middle ground on these design issues. The housing industry is a local-based industry. The developer–civil engineer–local government relationship is structured around an upcoming project. The property owner, developer, or contract purchaser retains the services of a land development engineer to design the project. The design plans are reviewed by other engineers in various government agencies. Land development engineers producing high-quality designs are always well received by regulatory agencies, owners, and contractors. By eliminating multiple reviews, the approval process becomes less complex and time consuming.

Land development engineers will continue to be challenged as regulations change, property characteristics become more difficult, and community influence becomes the norm.

Chapter 14

Developers, Engineers, Regulators

The art of bringing raw land to finished lots in a cost-effective and timely manner hinges on many factors, none more important than the relationship between the developer, engineer, and regulatory official. Regardless of role in the land development process, a project is better managed by understanding the perspectives of each team member. By appreciating different viewpoints, engineers will be able to coordinate the interests of each party resulting is a common goal—an approved project (Figure 14-1).

The engineer continues to be placed in the middle of diverging interests. It is commonly phrased that "land development engineers are between the sword and the wall." In this case, the sword is the client, and the wall is the government.

In this chapter, the traits of each team member are analyzed based on the following business considerations:

- Timing and schedule,
- Quality,
- Cooperation,
- Knowledge,
- Work experience,
- Commitment,
- Communication,
- Meetings experience,
- Documentation, and
- Government relations.

The land development industry has changed since 2008 with the disappearance of many regional or local land development companies. Perhaps in the future there will be more design-build approaches for residential projects. Regardless of the approach taken, land development engineers are facilitators, negotiators, and communicators with the intent

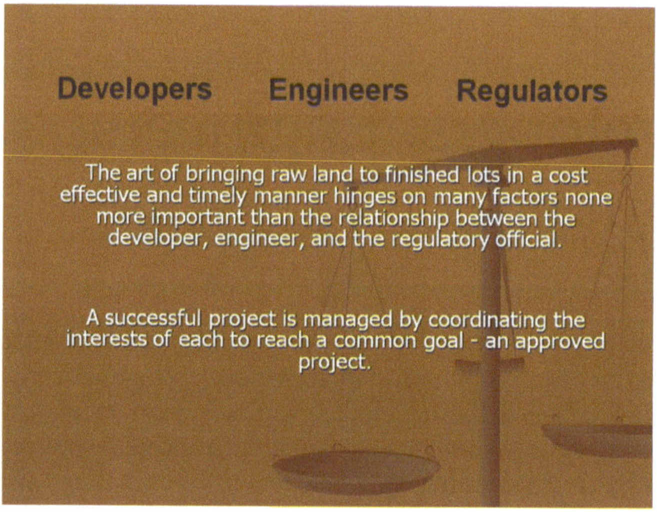

Figure 14-1. Land development engineers must rely on excellent interpersonal skills to orchestrate the cost-effective and timely design, permitting, construction, and infrastructure turnover of any residential project. The ability to communicate with a wide variety of people is a trait that lends itself well to the land development industry.
Source: David E. Johnson.

of having regulators and developers on the same page with a shared vision to enhance the community in which they all work.

Timing and Schedule

Developer

- Time is money, regardless of the stage the project is in.
- First, hurry up with the plans, then wait for changes based on marketing data.
- Last, hurry up to finish plans for submission.
- Owners always want to accelerate the schedule, even after a schedule has been agreed to by all parties, and even with a set regulatory schedule.
- Developers strive to accelerate the schedule during negotiations without understanding the business acumen of the consulting engineer.

Engineer

- Time is money.
- Land development engineers should not negotiate a schedule unless the client is willing to compensate the designer for accelerating the work, which may entail overtime or assigning additional personnel to the team.
- What does it tell you about your client if they seek a quicker turn around but are not willing to properly compensate the design firm?
- What happens *if* the accelerated schedule has been negotiated but missed by the land development engineer? Missing a deadline is unacceptable, but missing a deadline based on a negotiated schedule would be a business disaster. What do missed deadlines tell the client? The building industry is close-knit, and engineering reputations can be tarnished by missed deadlines.
- Less time to prepare construction drawings will typically result in a less-than-quality set of plans, which prolongs the project and associated costs.

Regulator

- Time *is not* money: the cost borne by the developer or engineer has no bearing on the regulatory schedule, and very few agencies will expedite a plan review or short circuit the permit process.
- Most regulatory agencies have a specified timetable to review and approve a set of construction plans. Even if government is late owing to the lack of staff and does not fulfill their statutory requirement, the staff will ask the developer to approve an extension of time. The bottom line is for the design engineer to submit a high-quality design and plan set.
- An expedited design schedule will most likely result in a subquality work product. Government reviewers will remember the subpar product and will be very cautious of subsequent plans submitted by the engineer. Reviews may be lengthened because of the lack of quality simply because government equates subpar plans to "probably" being wrong. The result is more time spent on the project by all the team.

Land Development Engineer's Goals

Understand the needs of the client by meeting with them, walking the property, and discussing the project with government reviewers. Meet specifically with the client decision maker and the specific person that will review the plans:

- Spend time creating a reasonable project schedule that meets the goal of the client and adheres to the regulatory review timetable. Obtain concurrence from the client and stay on schedule.
- Avoid negotiating time frames and schedules unless compensation is part of the discussion and your design team can deliver the project on time.
- Design an excellent quality set of construction plans is the ultimate answer to bringing all perspectives together.

Quality

Developer

- Expect quality from the design professionals without regard to the complexities of the site, regulatory environment, or owner changes.

Engineer

- Quality design and construction drawing preparation establishes a firm's reputation with developers and regulators.
- Engineering firms strive for a high level of quality, which will translate into faster regulatory reviews, minor comments, and fewer questions during the bid process and construction.
- Government can find issues with any set of plans simply because it is easier to review than to create. With available land being more difficult to develop and government regulations becoming more complex, the engineer finds delivering a high-quality product that much more difficult to achieve.
- In certain situations, engineers have used regulatory plan reviewers for their quality control, which is a double-edged sword. Land development engineers are in a continuous state of flux with the client and regulators.

- Constant changes and short time frames to make plan changes affects the quality of the plans. Clients make changes for many reasons, and the engineer must be cognizant of the schedule and how the changes will affect the schedule or quality of the plans.
- Engineers typically do not like design criteria changes or project approaches once design has commenced, because it becomes counterproductive and impacts the firm's workflow.

Regulator

- Nationally, the quality of construction drawings is about 5 out of 10. This is based on the survey of seminar attendees at the ASCE seminars on land development. The quality is based on plan presentation rather than design approaches.
- Review comments on design can be a result of many factors attributed to all three entities. For local government, the factors could be
  - Local government relying on outside consulting engineers to review project plans;
  - Multiple reviewers assigned to the same project; or
  - Reviewers who impose their individual influence beyond the framework of ordinances, standards, and regulations.

Land Development Engineer's Goals

- Know the local, regional, and state regulations better than the agency reviewers.
- Know the reviewers and establish a professional working relationship with each reviewer.
- Submit quality work on time.
- Revisit agreements and charge your client for multiple government reviews.
- Submit quality construction plans to minimize government comments, which
  - Expedites the review process,
  - Saves clients money and time,
  - Saves the engineer money and time, and
  - Encourages government to appreciate and remember high-quality plans.

Cooperation

Developer

- Expect cooperation, but successful developers know the balance between asking and demanding.
- Most developers start at the top levels of management and then work down through the organization to ensure the project remains on schedule.

Engineer

- Land development engineers rely on a strong sense of cooperation with all team members.
- Consulting engineers typically start at the bottom of a client's organization (project manager) and agency personnel (plan reviewer) to accomplish tasks in a timely manner.

Regulator

- Most regulators minimize regulatory delay and promote sound sustainable land development practices.
- Maintaining a strong sense of cooperation with all outside engineering firms and their clients will ultimately benefit the city as a whole.
- Some engineering reviewers prefer to only discuss project details with the consulting engineer and not the developer. In this case, consulting engineers need the latitude of making decisions on behalf of their client and minimizing any delays.

Land Development Engineer's Goals

- Always work in the best interests of the client by quickly responding to their requests.
- Base this industry on effective working relationships at all levels.
- Improve relationships with government staff; be aboveboard and ethically sound.
- Be proactive rather than reactive.
- Improve interpersonal skills.

Knowledge

Developer

- Most developers staff their teams with operational skilled professionals. Their abilities will be based on practicality and not necessarily engineering expertise.
- Developers will gravitate to the engineering firms most willing to help them succeed.

Engineer

- What type of knowledge is most important for a design engineer? Of course, technical knowledge would be the foundation, but knowing the regulatory process, land development regulations, and the plan reviewers would enhance civil engineering knowledge and expertise.
- Knowing the regulatory process is essential for preparing a reasonable timetable from application to construction start.
- Knowing the regulations better than the reviewers will be welcomed by the reviewers. By knowing local regulations, schedules will be improved and cost increases minimized.
- Knowing the reviewers (workload, trends, personal interpretations) is always be helpful.

Regulator

- What are their technical backgrounds?
- Do their backgrounds influence their review approach or how regulations are interpreted? Has there been a turnover of staff?
- Do they know their own regulations?
- Are the reviewers flexible or do they remain strict to the written regulation?
- What is the appeal process within the agency?
- What expertise or industry knowledge does the department director have?
- What is the department director's management style?

Land Development Engineer's Goals

- Know your clients and know their project goals.
- Know your regulators and know their hot buttons.
- Know the regulations and stay current on changing technology.

Workforce

Developer

- Are developers hiring engineering professionals to manage their field operations?
- Since staffing could be a problem in hot markets, developers should retain the services of a recruiting agency.
- From a developer's perspective, an active market requires more people because the demands are greater. In a slow market, a developer will have fewer people but the demands on them remain high. It is difficult to find qualified personnel.

Engineer

- Finding qualified personnel in active markets is difficult. Expect to search for people in other markets and pay for relocation.
- Retaining core group engineers during slow market periods is essential and crucial to the organization.

Regulator

- Government agencies are understaffed in both hot and slow markets.
- Engineers in the public sector have a greater sense of job security.
- It is difficult to find qualified personnel.

Land Development Engineer's Goals

- Companies are all in competition for people with land development experience. Land development firms would be served well by establishing an intern or co-op program with a civil engineering department. This would help establish a legitimate supply of engineers.

- Training and retaining new personnel are essential. It is better to train people and maybe lose them to the competition than not train at all.
- Elevate interpersonal skills as an important hiring criterion as even more important than technical knowledge.

Commitment

Developer

- Do developers care about what other projects engineers are working on? No.
- Should they? No.
- What happens if the engineer fails to meet their commitments? Future business consequences.
- What is a developer's commitment to his or her design professional? Future business repercussions.

Engineer

- Should engineers commit to a time frame for plan completion? Yes.
- Should engineers commit to a time when construction may start? No.
- Should engineers commit to saving the client money? No.
- Should engineers make special commitments to selected clients? Yes.
- Should engineers commit to quality designs and plans? Yes.
- Should engineers make commitments beyond their control? No.

Regulator

- Are regulators expected to commit anything to anybody? No, not really! However, personal integrity and the willingness to serve the community overrides any apathy.

Land Development Engineer's Goals

- Commitments made must be commitments kept.

- Land development businesses have always been based on relationships. If it is not a two-way commitment, strive to change the dynamics.

Communication

With today's ever-changing technological advancements in communications, it is much easier and faster to communicate and send plans than ever before. The problem is making sure all parties have updated devices and software to ensure seamless communications.

Developer

- Developers want to hear from their engineers on the status of their project. If they do not hear from them on a consistent basis, developers do not know if their engineers are working on their project.
- At the first client meeting, engineers and developers need to formalize the best way to communicate between the parties.

Engineer

- Engineers must keep all the pieces of a project in play and continually move projects forward.
- Government takes time to review project plans and after a submittal, engineers shift their priorities to other projects and issues. At this point in the process, the engineer must maintain a clear line of communication with the developer and with the municipality.
- Different clients may require the engineer to change the mode of communications based on the developer's operational plan.
- Regardless of communication technique, engineers must maintain a paper trail for future reference. Design decisions, submission dates, and discussions with regulatory officials must be recorded and distributed to all team members, especially the client. Phone calls should always be followed up with a confirming email documenting the actions or decisions reached during the discussion.
- Establishing a high level of customer service within the firm is vital. To the engineer, the developer and government staff are the "customers."

Regulator

- Every staff person will have his or her own way of communicating with engineers and developers. Also, depending on the agency, communications may be nonexistent, formal, or informal. Some regulatory reviewers do not want to hear from either the engineer or the developer.
- Most government officials do not want to meet or receive calls or emails from engineers and/or developers on a continuous basis, because it only diverts their attention away from their workload.
- Some regulators will only meet with the design engineer because the review is technical and not political.

Land Development Engineer's Goals

- Find a reason to contact your clients every week and provide a status report of their project. Providing reassurances that their project is a priority may be necessary.
- Prepare a *Memo to File* on important issues discussed over the phone or meetings with regulatory officials.
- Schedule regular face-to-face meetings with clients and regulators.
- Use all available communication tools and remain current on techniques.
- Act rather than react.

Meetings

Developer

- Most developers want structured weekly progress updates.

Engineer

- Design consultants should include pricing for meetings.

Regulator

- Regulators will meet to discuss projects and reviewed plans once properly scheduled.

Land Development Engineer's Goals

- Charge for weekly meetings (separate line item in the agreement).
- Manage the meetings (act and not react). Engineers should prepare an agenda and manage the meeting.
- Only essential personnel should attend meetings and be involved. Sheer numbers do not necessarily equate to better service.

Documentation

Developer

- Do developers conveniently forget directives?
- Do they remember only the problems?
- Are they always in survival mode?
- Do they have the staff needed? Limited administrative staff will limit documentation.

Engineer

- Document important milestones.
- Document project decisions and distribute.
- Summarize meeting discussions.
- Disclose how the schedule is affected by inaction by the client or government.
- Follow up to ensure all documentation has been received by those intended to receive documents.

Regulator

- Documentation can be lost, misplaced, or not recorded at all.

Land Development Engineer's Goals

- Projects are successful because of communication and essential documentation among the engineer, government agencies, and the client.
- Paper trails provide a foundation for additional services, appeals, and so forth.

Government Relations

Developer

- Government relationships are typically politically motivated.

Engineer

- The engineer balances client relations with government requirements.

Regulator

- Staff members enforce policies and regulations while maintaining a professional demeanor.

Land Development Engineer's Goals

- As development pressures continue, government at all levels reacts by adding regulations and constantly changing regulatory practices. The engineer must be a key participant at local community events and functions.
- Engineers must be the experts of choice in primary market areas.

Summary

1. Establish stronger relationships with government officials and staff reviewers. The relationship with government may be more important than client relationships.
2. Act rather than react.
3. Manage the client rather than be managed.
4. Do not negotiate time because time = quality = money.
5. Proactively communicate, consistently and constantly.
6. Document. The handshake relationship is no longer the business of choice, especially with national builders and the ever-changing face of regulations and regulators.

Chapter 15

Stormwater Management and Best Management Practices

Environmental impact of new residential housing developments has become a more important—and sometimes controversial—factor in gaining public support and regulatory approval. State and local governments are raising more questions about how stormwater runoff will be managed and how pollutants collected in the stormwater will be treated. Builder/developers and their engineers of choice must integrate plans for stormwater management (SWM) and best management practices (BMPs) for pollutant control into their designs for new projects.

This chapter reviews stormwater quality and quantity control facility options for residential development applications. This overall summary of current SWM practices provides methods to comply with local, regional, and state regulations. Then, each BMP suitable for use in a residential project is evaluated for application, cost, effectiveness, maintenance requirements, and public/private acceptance. The implementation of BMP standards varies among jurisdictions. Each has a preferred method for achieving water quality standards; also, they often have an established SWM approach and set facility design criteria.

To reach a sustainable solution, the design of each stormwater control facility must balance the application, cost, and environmental benefits. In planning effectively for stormwater facilities and water quality features, residential developers must evaluate property characteristics during the due-diligence period, rather than leaving them to be incorporated into the land plan later as afterthoughts. For example, a neotraditional housing project locates the garage, accessible through an alley, in back of the lot, with a public street in front. Also, the average home has increased in square footage, so the lot size may be smaller. The combination results in double frontage lots and with greater impervious surface area than conventional subdivisions. The water quantity and quality control approaches for these higher intensity uses become an issue in the design of user-friendly and cost-effective control facilities. Builder/developers should discuss these considerations with civil engineers early in the planning process.

SWM regulations have now evolved to require control of the amount of runoff discharged from the site to be equal or less than the discharge from predevelopment conditions. Design considerations vary greatly, from controlling the discharge based on a 100-year storm to controlling the discharge from only a 10-year storm event. The difference in required land area, construction costs, and maintenance requirements are significant issues for developers. Because the land development industry is fragmented and government regulations vary considerably, this chapter outlines the practical options available for residential applications. Developers should consider partnering with an engineering design team member and seek government change in subdivision design while meeting current water quantity and quality standards.

Decentralized versus Regional SWM Facilities

As development emerges along fringe areas of existing infrastructure and expands urban centers, local governments must assess the advantages and disadvantages of utilizing a regional SWM facility for future development. This forward thinking shifts the burden from the developer to a government entity, but the concept of regionalizing stormwater control has merit. Regional facilities are designed for specific watersheds, and the future development pattern and drainage analysis are based on the master plan and zoning map. The facility is designed, constructed, operated, and maintained by the municipality.

The results of this approach are significant in many respects. Monitoring one facility rather than multiple operations is obviously less costly. Regional facilities also remove the operation and maintenance responsibilities from multiple entities in cases where the jurisdiction does not accept responsibility for maintaining the facilities. In most instances, decentralized water quantity control measures are not maintained properly, and the facilities are not inspected annually. Centralizing SWM facilities is consistent with how regional wastewater treatment plants serve large areas. The projects upstream of the SWM facility pay for conveying their discharge to the facility; fees are based on the difference between pre- and post-development conditions. Usually an annual maintenance fee is required for each new project. Typically, homeowners associations (HOAs) include this fee in their annual budget or it is assessed against each lot and run with the land rather than the ownership entity.

In many HOAs, the covenants, conditions, and restrictions (CCRs) and other association documents are not transferred to new owners upon resale. With a regional SWM facility, the problem of transferring responsibility is eliminated. This regional facility could be planned within a park system. In contrast, decentralized SWM facilities are often placed in less noticeable areas and may be difficult to access to properly monitor operation and maintenance. Although facility landscaping, fencing, and side slopes may be adequately maintained, the control structure may not be routinely checked.

Communities wishing to enact a regional approach to SWM may need to change their ordinances to increase open space requirements for passive or active recreational use. This way, uplands are not used to control stormwater runoff, taxable land area increases, and the perceived value of the project and home sites also increases. The cost of operating and maintaining the regional facility is paid by the users within the watershed. The government's review process of SWM engineering designs, which is now long and difficult, must also change. Regulatory review proceedings need to be shortened and costs shifted from hard construction expenses paid by the builder to direct known payments for regional facility connection. Builders prefer direct payment since it is a known cost incorporated into the project financials and does not change over the build-out period. Homeowners prefer this arrangement because the operation and maintenance of one or more SWM basins is not their responsibility. HOA liability insurance policy also changes. Finally, the regulations governing the quantity and quality of stormwater runoff affect the way property is developed and significantly affect the cost of development. Because each property and development design program differs, engineers should have flexibility in stormwater design so as to achieve sustainable projects.

A disadvantage of decentralized SWM facilities has to do with the timing and rate at which stormwater is discharged. The peak flow from a well-run, properly maintained, and decentralized SWM facility could still combine with the discharge from other decentralized facilities to have a detrimental effect on areas downstream. Whatever the SWM facility—whether regional or decentralized—it will alter the characteristics of the surrounding watershed and must be evaluated for the impact it will have on downstream properties.

Stormwater Management

Retention Facilities

Retention facilities are basins or ponds that retain runoff from a site and have a permanent water surface (Figure 15-1). The basins are designed with overflow control structures and emergency overflow weirs. Although basins have a constant water surface, the stormwater runoff generated from post-development conditions is stored above the permanent water surface elevation.

Areas with site conditions conducive to permanent water surfaces should be fully utilized in land planning. A retention facility serves

Figure 15-1. Retention basin. Stormwater management facilities with a permanent water surface are called retention basins. If designed properly, the basin is a designed amenity for the project, including a walking path around the basin.
Source: David E. Johnson.

multiple roles. Excavating the basin to create a pond generates fill material that can be used on site. The pond provides the necessary stormwater quantity control, but also contributes to meeting water quality standards. The depth of a retention basin is a design concern for the steepness of the side slopes and the necessity of benching the slopes to the bottom of the pond.

Incorporating the basin into a land plan enhances the marketability of the project and adds value to homes. The pond also provides quality open space and expansive views. Because of these benefits, builders should consider retention basins as their first design choice. In many residential applications, a pond, lake, or retention facility is considered a desirable view and builders can assign premiums to the adjacent lots. Buffer space along the retention facility may be necessary to control activity and provide an additional water quality measure.

Detention Facilities

A detention facility controls runoff by holding it for a period of time, then releasing the stormwater at a controlled rate (Figure 15-2). Detaining stormwater runoff is the standard design approach of most engineers and regulatory agencies. Detention facility designs can be above or below ground and can be adapted easily to the land plan and site conditions.

Land assessed for development potential may have site constraints, including significant terrain issues that result in multiple drainage areas and multiple SWM facilities; detention basins are typically used in these situations, because basin sizes can vary and be graded into the site's natural features. Although detention basins are considered open space elements, they do not necessarily add to the value of homes nor are they considered aesthetically pleasing. The detention basin is designed to store site runoff, detain it, and discharge it at the predevelopment rate. Thus, much of the time, the detention basin is dry. Most basins have an underdrain system for drainage. Several issues must be considered in designing a detention facility.

The detention facility may require that a berm be graded around the basin, especially if it is located in sloped terrain. The berm must have a minimum width of 10 ft for maintenance access. The berm can be designed with an emergency weir, which may be grass or concrete. The side slopes outside of the basin should be designed for ease of maintenance. The depth of the basin is also a design issue and may require gradual slopes for safety and maintenance considerations.

Figure 15-2. Detention basin. Stormwater management facilities that are dry most of the time are called detention basins. Municipalities require fencing and landscaping if basins are visible from the interior subdivision streets.
Source: David E. Johnson.

The location of detention facilities within the project site is critical for design efficiency. The location also affects construction and maintenance costs. Locating basins in steep slopes, unsuitable soil conditions, or in "leftover" project spaces may seem appropriate to maintain lot yield while maximizing use of the land, but the results may not be cost effective or achieve the requirements of the municipality. In most cases, detention facilities have limited capacity and must be supplemented with other measures.

The control structure for a detention facility is typically a corrugated aluminum riser pipe or a concrete structure. The riser has multiple outlets discharging the flow at controlled rates. It is typical to design a 3 in. orifice in the structure to allow for low-flow discharge. A detention facility may also have weirs or overflow capabilities to pass through off-site drainage or large storm events. The control structure should have trash racks installed to ensure the facility functions as intended and will not become clogged by debris. Many detention basins require a low-flow channel along the bottom to pass low flows from the pipes' discharge locations into the facility and direct the flow to the control structure. The low-flow channel is typically concrete but may be constructed of pavers, interlocking blocks, or grasscrete.

High-intensity uses such as commercial, office, or multifamily housing projects may have high raw land prices, and using valuable land for SWM facilities may not make practical or financial sense. A reasonable alternative is to design underground detention facilities, located under parking lots or in open space areas. Typical designs detain site runoff in underground pipe systems and infiltration beds, allowing it to be caught in the pipe system and discharged through soil infiltration and via a control structure system discharging to a positive outlet. Infiltration pipe systems may also be designed to control roof runoff by collecting rain leaders piped directly to the underground system. A wide variety of approaches are available for detention facility design, and the engineer must be creative in implementing a solution suitable for the site and project goals.

Best Management Practices

All levels of government have asserted efforts to control and manage the increased volume of land development runoff and reduce the runoff's pollutants. The increased runoff is a direct result of increased impervious areas associated with nonresidential and residential projects. Improving the quality of the runoff discharged from a developed site further protects the receiving waters, including ponds, lakes, streams, and rivers. When control of the flow discharged from a developed site is integrated with BMPs, the quality of receiving waters also improves. Although many design alternatives are available to improve water quality through the implementation of BMPs, changing subdivision design standards and reducing impervious surfaces are also options.

Many jurisdictions have implemented new land development standards reducing impervious surfaces in the development of residential housing projects. They require additional open space and buffers, but they also evaluate road widths, parking requirements, and alternative driveway paving materials. Reducing roadway widths not only reduces impervious surface area but achieves lower vehicle speeds in a residential setting. This traffic calming approach is a design trend leading to project sustainability.

Roads are also changing from the typical curb-and-gutter section to a roadside swale section. This design approach was always acceptable in rural areas and was used in residential design years ago. The buying public has long accepted curb and gutter with sidewalks as a standard subdivision design approach; anything less is now considered substandard. The market will slowly change and learn to accept swale section roadways, perhaps without sidewalks. The benefits of using roadside swales to drain lots and roads are substantial in improving the quality of runoff from developed areas.

For creative design, the quantity of runoff could be controlled at each driveway culvert, a practice that also promotes water quality measures by infiltration. For this to succeed, however, governments must be willing to change, builders must be willing to accept the change, and the buying public must understand the benefits. If all builders comply with a new standard, they are on a level playing field and the public does not have a choice.

Providing more open space by designing reduced lot sizes, narrower roadways without on-street parking, and right-of-way swales, when combined with proper utilization of the land, achieves marketable projects while improving water quality and community sustainability. SWM and BMP facilities must be incorporated in the design program from the beginning, at the conceptual land planning stage. BMP design approaches include alternative land planning techniques requiring public and private leadership and design concessions. Water quality is achieved through infiltration, filtration, and retention and detention of runoff from developed areas. This section addresses only a limited number of design options, but they are typical BMP solutions for residential applications.

Bioretention Systems

In residential applications, the bioretention system is typically designed for small drainage areas and, in most cases, individual lots. These areas are frequently referred to as rain gardens (Figure 15-3). A rain garden must be located in the low area of a lot. The rain garden is shallow and installed after the lot areas have been stabilized; this reduces silt build-up within the facility and maintains operational integrity. The bioretention system consists of a soil mixture with an infiltration rate of 1 to 1.5 in/h in the bottom of the facility. This soil layer varies from 2 to 4 ft in depth and acts as a filter bed. An underdrain system is designed below the replacement soil to convey flow to an acceptable outlet; it has a

Figure 15-3. Rain garden. A bioretention facility or rain garden can be designed in many forms and function as a water quality device. Rain gardens consist of filter media and landscape material and are best designed for small drainage areas.
Source: David E. Johnson.

clean-out port for maintenance purposes. The rain garden is planted with trees and shrubs native to the area. A 3 to 4 in. layer of mulch covers the facility. Proper, regular maintenance of the rain garden is essential to ensure it functions properly.

A bioretention facility can achieve a removal rate of 85% to 90% of the total suspended solids (TSS) and provide decentralized water quality control, but the level of maintenance increases. HOAs and individual lot owners must be educated on these facilities so that they respect the integrity of the system through proper maintenance.

Roof Drywell Systems

In most residential applications, water from the back of lots and from rear portions of the dwelling's roof area is the most difficult to capture and treat to meet water quality standards. Most land planners run lots to the perimeter property lines and do not incorporate a buffer between the property line and the rear lot line. Installing designed drywells at each rain leader provides infiltration of the runoff, which reduces the overall runoff from the back of the lots; thus, the proposed water quality systems treat a smaller drainage area and less impervious surface.

The roof drywell system requires soil conditions with a permeability rate between 0.2 and 0.5 in/h. The roof rain leader connects to a perforated PVC pipe system encased in a stone fill. This design must include an overflow pipe and inspection port. The size of the drywell system is contingent on draining the runoff volume within a 72-h period; these facilities cannot be located in compacted fill areas or close to basement walls. The system design is vertical or horizontal with a minimum 2 ft separation between the facility bottom and groundwater. A maintenance plan is necessary to ensure the system functions properly. Designing a sump or trap in the system between the building and drywell minimizes particulates and sediment that enter the system and ultimately clog it. Drywell systems are best suited for areas with permeable soils.

Extended Detention Facilities

For many years, engineers designed detention basins to control the quantity of subdivision runoff and discharge it at a rate less than that of predevelopment. For water quality purposes, these detention basins are designed as multistage facilities (Figure 15-4).

The design intent is to detain the runoff generated from a 1- or 2-year storm event for a longer period. Extended detention facilities remove 40% to 60% total TSS; the longer the detention time, the more the TSS removal rate increases. An extended detention facility discharges runoff through a control structure and infiltration system; thus, the facility must remain dry between storm events. Although additional BMPs may be needed to attain the required TSS removal, this type of facility is typical for controlling a site's runoff quantity.

An extended detention facility is a dry basin with sufficient storage and an outlet structure designed to contain and slowly discharge runoff over a period of time. Basins are also designed with an overflow system that discharges into an acceptable downstream drainage system. Many designers strive for a shallow basin with a large bottom surface area rather than a deep but small basin. High land values or high-intensity land uses may necessitate extended detention basins designed belowground. Design considerations in these instances include depth of

Figure 15-4. Extended detention facility. This facility consists of a forebay, rain garden, and an extended detention facility. Stormwater runoff is controlled in an extended detention facility for 48 h before being discharged.
Source: David E. Johnson.

groundwater, soil permeability, a positive discharge point, and overflow capabilities.

Whether aboveground or belowground, extended detention facilities require a considerable maintenance program that includes maintaining the areas draining to the facility, as well as the basin components. Removing sediment from upstream systems ensures the basin will function properly. The maintenance program should be established for regular intervals—weekly, if necessary, to remove debris and sediment from the basin bottom, low-flow channels, and control structures.

Manufactured Treatment Systems

A manufactured treatment system consists of a prefabricated structure designed as part of the drainage system and located in the system prior to an extended detention facility. The combination of the manufactured treatment structure and an extended detention facility provides sufficient

TSS removal for regulatory compliance; several devices on the market remove up to 60% of TSS. A device in line with the drainage system is easily monitored for maintenance; it does not detract from the appearance of the project, nor does it require dedicated land, as it is located within a public right-of-way. The manufacturers are required to present their test data to each state agency for verification and certification.

The size of the manufactured device is based on actual discharge in the drainage system. Once the size is determined, the manufacturer can easily provide the developer with an installation price. The developer compares this cost to that of land dedicated to a different BMP method. To lower the cost of the facility, the device can be designed off-line. This allows the device to be designed for storm events that exceed a given threshold for water quality, and runoff from larger storms can be redirected past the device to an acceptable SWM facility.

Manufacturers provide maintenance programs to ensure the facility operates efficiently and as intended. The ownership of the system must be determined before the project design is started. However, many jurisdictions will not approve a manufactured device, and others will not accept the ownership even if it is located within a public right-of-way. Many residential projects require HOAs, so the maintenance of these systems is the responsibility of homeowners. Water quality treatment technology continues to advance, and manufactured treatment systems will be more readily accepted with proven testing and certification procedures.

Sand Filters

A sand filter consists of a forebay, sand filter bed, underdrain system, and overflow device. The *forebay* is a storage area for removal of debris and sediment before the flow enters the sand filter. In essence, the forebay acts like a detention basin with an overflow device; it filters the runoff before discharging into the sand filter. The forebay also has a drainage system to empty any standing water for maintenance purposes; this prolongs the life of the sand filter. The TSS removal rate ranges from 80% to 85%. Located aboveground or belowground, sand filters are best suited for nonresidential areas and high-density residential projects. The sand filter thickness and permeability must be sufficient to remove pollutants and drain the facility within a specified period, for example, 72 h. Sand filters are maintained, including sediment and debris removal, after every storm that exceeds the water quality design. The sand material is tested annually to confirm the permeability of the sand bed. If water in the sand filter does not sift through during the design time, the sand bed is replaced. Thus, a diligent maintenance plan must be implemented and followed without variation. Sand filters are expensive, and aboveground filters are usually located in upland areas that have suitable soil characteristics. Sand filters are typically controlled by the property owner in a nonresidential application.

Vegetative Filters

Vegetative filters are a BMP option when buffers are used to enhance the layout through open space elements and retention of native vegetation. These linear buffer areas remove pollutants from stormwater runoff sheet flow rather than concentrated flow. The total TSS removal rate varies depending on the vegetative cover, but a typical removal rating is between 60% and 80%. Design choices for a vegetative filter include uniform graded slopes, width of the filter area, slope of the area contributing to the filter, type of vegetative cover, and soil characteristics of the site. The width of the vegetative filter varies with the slope of the contributing drainage area; most jurisdictions limit the slope to a maximum of 8%. The cost of a vegetative filter is prohibitive when the buffer is planted with new vegetable and groundcover material. However, if the vegetative filter is incorporated in the land plan, the cost of saving natural vegetation is offset by the marketability of the lots or homes bordering the filter. Healthy vegetation is important for the system to function properly; regular maintenance inspection of filter strips must be undertaken by the HOA. The filter strips or buffer areas must be contained within a conservation easement or a dedicated open area parcel of land.

Permanent Water Surface Ponds

Permanent water surface ponds, also known as wet ponds, can be designed to control stormwater runoff and improve water quality (Figure 15-5). With a properly designed control structure, this pond acts as an SWM facility, storing runoff above the permanent water surface elevation and releasing it at a rate comparable to that of predevelopment drainage. Wet ponds are also the most effective means of removing pollutants generated by land development projects. Ponds acting as BMP facilities can be designed within a housing project as premium open space. Most builders use the wet pond as a focal point of a subdivision and integrate the feature into the aesthetic presentation.

Figure 15-5. Permanent water surface. A pond or lake with a permanent water surface is designed with a control structure and concrete emergency overflow weir. The post-development runoff is stored within the water body above the permanent water surface elevation and discharged at a controlled rate.
Source: David E. Johnson.

The cost of constructing a wet pond is in grading and the control structure. The pond and control structure design is important to the overall performance of the facility; the project must have sufficient land area and topography characteristics to support a wet pond for water quality purposes. Maintenance requirements include removing sediment and debris, cleaning inlet and outlet structures to ensure full operation, and controlling algae. In Florida, for example, wet ponds are common because the water table is so high; they provide fill material to generate site grades, control site runoff quantity, and provide a high-efficiency TSS rating for pollutant removal from runoff.

Soil Testing Programs

For SWM and BMP facility design, the builder/developer must undertake an extensive soil testing program for the site. Using the base map and project land plan, the location of the water quality and quantity control systems can be identified, and testing should be undertaken early in the design process. The results of the soil testing indicate the type of soil underneath the SWM facility; soil permeability is important in calculating the detention basin's drain time.

This is the case for underground detention facilities and permeable asphalt. Soil samples must be taken at varying depths for each facility. Extensive testing of the soil at SWM facility locations provides important data that can be used to design efficient and cost-effective facilities.

Visual Considerations

Some local governments impose various requirements as to the appearance of SWM facilities. These requirements may pertain to fencing, access gates, or all-weather access ramps for safety and maintenance purposes. In many jurisdictions, detention facilities are landscaped to become part of the open space rather than a disjointed hardscape feature.

Landscaping

In some instances, builders are required to provide substantial landscaping to hide SWM facilities from view (Figure 15-6). The intent is to transform a dry detention facility into a landscaped feature. In many cases, however, the aesthetic results do not justify the cost of landscaping material; moreover, the landscaping material usually is not maintained properly, especially in single-family home subdivisions with HOAs. Overgrown or excessively landscaped detention facilities are detractions and eyesores. A balance must be sought between achieving an aesthetically pleasing feature for the initial opening of the project and long-term appearance. The selection of ground cover for the basin may also be dictated by the municipality. As many detention facilities are beyond irrigation coverage, the selection of drought-resistant plant material may be necessary. It is not uncommon for builders to retain the services of a landscape architect for the design, material selection, and details. The side slopes within the basin are sometimes planted with native grasses to eliminate or reduce maintenance.

Figure 15-6. Landscaping material. Stormwater management facilities and best management practices require landscape material designed to enhance runoff filtration. In most cases, landscape architects are involved in selecting the proper plant material.
Source: David E. Johnson.

Fencing

Most detention facilities must be fenced as a condition of design approval. Builders use a variety of fencing style and materials. The two most common approaches are black vinyl chain link and split rail fencing. Black vinyl chain link fencing is less noticeable than landscape material, provides longevity, and is cost effective. Split rail fencing typically has wire mesh secured to the inside of the fence. Builders select the fence material based on the location of the detention basin.

Maintenance Access

Most municipalities require accessibility to the detention basin for maintenance, including a 6 ft swing gate for maintenance equipment.

The maintenance ramp into the detention basin should not exceed a 10% grade and is typically constructed of all-weather material. The side slopes of the basin should not be steeper than 3:1.

Maintenance Programs

Local jurisdictions require SWM facility maintenance programs to be submitted with the project's final construction drawings. These programs include the entity responsible for the operation and maintenance of the facilities (Figure 15-7). The maintenance program also delineates the tasks required for weekly, quarterly, and annual maintenance.

In general, weekly maintenance is limited to mowing and debris removal. Quarterly maintenance is for inspection of the facility to ensure proper operation, remove debris, and repair fencing. Annual maintenance includes silt removal, facility inspection, plant removal and replacement, seeding, and general repairs. Most detention facilities do

Figure 15-7. Maintenance. Most stormwater management and best management practices facilities are owned and maintained by HOAs. Specific maintenance programs are included in the HOA documents.
Source: David E. Johnson.

not require extensive or intensive maintenance, but if the general conditions of the street and drainage system are not properly maintained, the basin will also need attention. To enhance maintenance of SWM facilities, designers typically design 4 in. perforated PVC underdrains within the basin connected directly to the control structure. For residential subdivisions, HOAs should budget for each basin's maintenance cost.

Ownership Alternatives

Most local governments require water quantity and quality control facilities, but they may not accept the ownership, liability, or maintenance responsibilities. The design and property encumbrances are the same for privately or publicly owned facilities. Typically, one of three entities is charged with the ownership and maintenance of SWM and BMP facilities: government, HOA, or property owner.

Government

Local government can own and maintain SWM and BMP facilities. The jurisdiction may require that the developer post sufficient monies to ensure perpetual funds for the maintenance of specific project facilities. In other cases, the local jurisdiction may require the developer to maintain the facilities for a 10-year period, at which time the public works department accepts it for maintenance. This 10-year period is bonded by the developer, who may be required to pay inspection fees up front.

The ownership and maintenance of many disjointed and isolated facilities, small or large, have caused local governments to rethink their approaches. In some cases, government requires the developer to create an entity that owns and maintains the facilities in perpetuity. In other instances, however, progressive communities have enacted master plans that manage SWM and BMP facilities to control watersheds rather than subareas within the watershed. A drainage district may be created so that facilities can be operated and maintained like wastewater treatment plants or public rights-of-way. In this instance, the developer contributes to the watershed fund based on impervious surface percentage or difference in pre- and post-development conditions.

Homeowners Association

In single-family detached subdivisions, SWM and BMP facilities are located on separate parcels of land commonly controlled by the lot owners. The developer must establish a legal entity to own, manage, and maintain the property. Many jurisdictions require open space areas, amenity buildings, and recreation areas deeded and owned by the association; in these cases, the HOA should also be charged with maintaining SWM and BMP facilities. Most home buyers are accustomed to having an HOA or umbrella community association to own, manage, and maintain common area facilities, including those controlling the quantity and quality of the site's stormwater runoff. The HOA budgets for the tasks necessary to ensure proper operation of each facility. Because these facilities are situated on separate parcels of land owned by the association, property taxes will be assessed. The larger the project and number of homes, the greater the chance the HOA will remain a viable entity and carry out its responsibilities. In subdivisions with large lots, rural areas, or a limited number of homes, it is possible the association will not actively engage in its responsibilities, and this problem is further accentuated after each resale and turnover in property ownership. Documents are not always transferred, and new owners are not always aware of their responsibilities—they find out only when the municipality issues a special assessment against their properties.

Individual Property Owner

The third option available to developers is locating an SWM facility on an individual lot where it is contained within a drainage easement. This is advantageous in locations where creating a controlling association is more detrimental in the marketplace than encumbering one lot with the responsibility of owning and maintaining the SWM or BMP facility. This approach may work for multiple homeowners if the facility is large enough to span several lots; however, developers should strive to restrict this encumbrance to only one lot. In many cases, a dry detention facility requires mowing, weeding, and general cleanup—not much different from regular property maintenance. Home buyers purchase the house with the facility with full disclosure and understanding of their responsibilities. The property is deed-restricted and maintenance responsibilities are clearly delineated. The homeowner may be eligible for property tax relief on the area associated with the SWM or BMP facility.

Many communities now impose BMP approaches that control every drainage subarea of a residential lot; this is burdensome when lots back up to adjacent properties, other lots, or drainage conveyances, such as flood plains. Individual lot BMPs are designed, easements created, and deed restrictions established. The idea of a homeowner properly maintaining a 200 ft^2 rain garden may seem unrealistic; it might work for the initial homeowner, but the concept may fade with each resale.

Design Trends

Controlling the quantity and quality of runoff generated from developed areas continues to evolve as more test cases are studied and results are published. Municipalities are following the direction of state environmental agencies and are changing design parameters to fit local conditions. Many state agencies do not agree on pollutant removal levels; there is a need for a national clearinghouse on SWM and BMP information that can be disseminated throughout the engineering industry. Consistency in design criteria, use of standard details, and results expected for each method are needed to ensure proper and expedited government reviews.

Chapter 16

Permits and More Permits

The housing industry is one of the most highly regulated industries, and the regulation comes from all levels of government. Each housing project must obtain various permits and approvals from local, state, regional, and federal government agencies before construction can begin (Figure 16-1). The housing industry is the only industry in which the community and the public as a whole can influence the design and approval of a project. The public, by objecting to a proposed residential project, can affect the approval process and the business success of a development firm.

Many municipal planning boards are composed of political appointees, who should cater to the public. It is the position of planning boards to achieve a consensus of opinion before a project can be approved. The developer's design team must present the positive attributes of the project to persuade the board to vote affirmatively for the project. In general, large projects gain more public attention, and with that attention more objections will be raised—founded or unfounded. The planning board members must objectively review both sides of the case and decide what is best for the community. Unfortunately, those in need of a new home usually become knowledgeable about a project only after the sales office opens for business. During the regulatory process, the buyers in the housing market do not know of upcoming projects and therefore miss the opportunity to voice their opinions.

As the no-growth environment continues to spread and regulations expand, it will become more difficult for large housing projects to gain approvals. The cost for the design team will be substantial. The effect will be to encourage the development of smaller multiple sites in noncompeting locations to satisfy different segments of the housing market. A successful approval process and planning board hearing are contingent on the knowledge and persuasive capabilities of the design team. Their knowledge of the regulations implemented to govern land use and development ordinances must be impeccable. Builder/developers, however, cannot rely solely on the design team to look out for their best interests. The owner or delegated staff members must become more aware of the

Figure 16-1. Permits. Residential housing projects are subjected to multiple government reviews and permit approvals, including the approval of water and sewer extensions serving new subdivisions. Local, county, regional, state, and federal government agencies are involved in the approval and permitting of residential projects.
Source: Sidney T. Burns. Due Diligence Associates, Parkland, Florida.

regulations that affect the building community. In the development business, they must guide and lead the design team to an all-win attitude at the planning table.

The regulations implementing planning and zoning ordinances change through the interpretation of those regulations by staff members. It is this interpretation that causes an uneven playing field for those developers competing in the same marketplace. If the preferences and interpretations differ from reviewer to reviewer, the regulatory review and approval will differ from project to project.

As additional regulations are implemented, the cost of land development will increase. Most communities act on the side of conservatism, which translates into higher costs. If the governmental reviews are not consistent with the regulations, an imbalance in the housing marketplace ultimately causes the approval process to be governed by politics. This chapter examines the various layers of government that are involved in the permitting and approval process and recommends ways to effectively manage planning board hearings. It is important to note that projects successfully permitted in a timely manner are a direct result of the quality of the design plans and "like-minded" concepts among government, developers, and engineers.

Local Governments

Local governments favor housing projects consistent with their master plans and land use regulations. A proposed housing project consistent with the zoning ordinance is reviewed by the planning board. The board of adjustment acts on projects seeking relief from zoning regulations. Designing the project so that design waivers and variances are not needed increases chances of success at the planning board. An excellent trial balloon for a project is the conceptual site plan review process, which is an open exchange of views among the developer, regulatory staff, and politicians. The views are not binding for any party. The preliminary and final plan reviews, however, affect the project design parameters. All projects should be presented in a positive manner knowing that the project will not become a reality until the final subdivision map or plat has been approved for recordation.

Planning Board

The planning board is the primary legal government agency that manages the development process. All other agencies support the planning board and are charged to provide specialized expertise in reviewing subdivision proposals. Each jurisdiction has various community interest groups influencing the subdivision process. The various agencies that review the project plans and determine the effect the project will have on public services are considered next.

Shade Tree Commission

This citizen group reviews housing projects for their effect on existing trees. Shade tree commissions are charged with managing and protecting existing trees and enhancing community beautification. This commission also reviews housing projects for compliance with landscaping, tree preservation, and reforestation regulations.

Environmental Commission

Many local governments have created environmental commissions to review design plans for endangered species, endangered vegetation, and wetlands. All environmental concerns that may affect the project are duly considered during the review process. In most cases, the environmental commission consists of politically appointed people having a sincere interest in protecting the environment from possibly destructive construction of new housing projects. The local environmental movement becomes much more vocal and much more intense during the review process. The environmental commission highlights problem areas and leaves the solutions up to the design team. Developers should welcome input from the commission and use it to find ways for subdivision designs to complement the environment (Figure 16-2).

Water and Sewer Departments

Community water and sewer facilities fall under the jurisdiction of either a separate municipal utility authority or the public works department (Figure 16-3). The design of water and sewer facilities for a residential subdivision is reviewed by regulatory engineers at the local and state levels. In many cases, the utility design review has a separate time schedule from the planning board process. As long as water and sewer are available, jurisdictions will approve a subdivision, subject to the utilities being approved by the appropriate government agency. In some communities, the local utility design is subject to county, regional, and state approvals.

Drainage Departments

Storm drainage facilities for residential subdivisions are usually reviewed and approved by the public works department; in some cases an outside consulting engineering firm is retained to review the engineering design work. Implementation of the local drainage ordinance varies from one engineer to the next. Knowing the reviewing firm and its preferences in drainage design is important for achieving approval during the initial plan submission (Figure 16-4).

Figure 16-2. Subdivision design. Community sustainability requires compromises in design standards and government permitting requirements. New land uses and difficult land forms provide opportunities for creative subdivision design.
Source: David E. Johnson.

Traffic Departments

Communities may also have traffic districts or transportation districts to evaluate the effect of the traffic generated by the proposed subdivision (Figure 16-5). This agency concentrates on adjacent existing roadways and intersections. In many growth areas, local government strives to implement taxing districts or assess impact fees against the housing project for its fair share of public improvement projects. Unfortunately, the assessment of fees and the transfer of community responsibility onto the new home buyer have resulted in higher prices for homes in those areas.

Project Reviews

The project improvement plans could be reviewed by up to seven consulting engineers who have varying backgrounds and represent all

Figure 16-3. Water and sewer. The availability and capacity of existing water and sewer facilities continue to guide residential growth patterns. Infill or overlooked properties are developed cost effectively because of the availability of existing infrastructure.
Source: David E. Johnson.

levels of government. These engineers review, approve, reject, and/or revise the design intent of the project.

Once a final plat is approved, municipal governments may require a performance bond or a letter of credit for the subdivision to be posted at 120% of the improvement cost. Also, an inspection fee may be assessed so that the consulting engineering firm representing the government can inspect the project during construction and during the maintenance bond period. For any subdivision, regardless of the size, the land plan should be evaluated for phasing. Phasing will conserve the cash flow to build the project and achieve maintenance bond status in a shorter period of time. The primary objective of land development operations is to build the improvements with high-quality workmanship to gain maintenance bond status as quickly as possible. This approach limits the project liability for the infrastructure and begins to establish responsibility for the operation and maintenance of the infrastructure with local government.

Figure 16-4. Alternative technology. New regulations are the basis for new design approaches. However, there are limited technological advances in infrastructure component design and methodologies owing to the fragmentation of the industry. Use of alternative pipe material for water, sewer, and drainage facilities is readily accepted at the local level.
Source: David E. Johnson.

Most municipal engineers prepare bond estimates using inflated unit prices. In general, these costs are much higher than private industry prices established during the bid process. Serious discussions should be initiated if the bond estimates are extremely high. Limiting the problems at the performance bond stage will help the overall corporate financial position on the project. A favorable bond estimate may assist in using the additional financial resources or bonding limits for subsequent project phases or housing projects.

Having several small projects at different levels of approval or construction permits the business cash flow to be properly managed. The phasing of projects in areas that are not governed by a regional agency provides an easier path to the building and selling of lots and houses. Locating projects within several political jurisdictions with varying approval processes is a business plan designed to earn a profit, generate

Figure 16-5. Subdivisions for vehicles. This wide entrance road is softened by the bridge treatment and paving materials. Road design for residential applications must complement pedestrian traffic and provide traffic calming devices created to reduce vehicle speeds. Source: David E. Johnson.

business, establish an employment base, and maintain a corporate presence in the community. Having several small projects reduces the risk factor, and the business growth goals can be achieved, even if one project is delayed.

Regional Agencies

Regional Planning Boards

Cities or incorporated towns may have their own planning boards. In addition, there may be a county planning board, which oversees county-owned roadways but also approves the subdivision plan. This layer of government review is not uncommon, and each housing project is subjected to another review by engineers and another round of political scrutiny.

Soil Conservation District

The soil conservation district in most states is a regional agency that governs the review and approval of sedimentation and erosion control design measures for residential projects. This agency governs all soil disturbance activities. Residential project submission and review fees sustain the district's budget. The soil conservation district's review process has a separate regulatory approval time frame. It is common for the regulations of the soil conservation district to be in conflict and expand the regulations imposed by local, regional, and state agencies. At times, a developer may be caught in the middle between agencies. This regulatory conflict will delay the approval of the project unless the art of negotiation and compromise can be used with the different agencies. This situation is damaging to a project schedule, and direct management of the situation is required by the builder/developer.

Stormwater or River Basin Agencies

Some states have regional stormwater or river basin agencies responsible for reviewing residential storm drainage designs to determine the effect the project will have on major streams, rivers, or other bodies of water. Review fees, application fees, and the approval process are separate from the local approval procedure. In some states, stormwater management facilities may be regulated by four to five government agencies. The duplicated, conflicting, and excessive regulations governing residential storm drainage design add to the cost of homes and delays the approval process.

Protection, Enhancement, and Proper Use Groups

Some states have regional groups established through legislation that govern large areas for protection, enhancement, and proper use. This regional agency structure is layered over other local, regional, and state approvals. As an example, in New Jersey, the Pinelands Commission is responsible for the overall development and land use of the National Pinelands Area. The Pinelands Commission approval process is complex, time consuming, and very expensive. The Pinelands Commission has its own regulations, which in most cases are much more restrictive than other agencies involved in the overseeing of residential development.

Also, in New Jersey as well as in other states, there are agencies that review housing developments located near the ocean.

Quasi-Public Agencies

Other regional agencies are quasi-public, such as the phone, electric, gas, and cable companies. These companies also review the construction drawings of the subdivision. The underground utility layout will be based on the engineering design plans. In some cases, the utility design does not complement or cannot be coordinated with the design of the subdivision. The utility company may locate structures along common property lines at the right-of-way line. Thus, the project engineer should evaluate the location of driveways, storm drain inlets, manholes, and yard swales that may be in conflict with utility company transformers, cable boxes, telephone poles, and so forth. If there should be a conflict, revise the engineering plans to complement the utility company design for the project. If thc change can be made without disrupting the project, change the plans. Changing the utility company's design will take longer to accomplish.

State Agencies

States oversee the implementation of housing master plans within the political subdivisions of the state and evaluate the effect housing has on the state's natural resources. States emphasize the protection of the environmental attributes of the state, such as wetlands. Some states have exceeded federally mandated wetlands guidelines and expanded their role in the protection of wetlands within state boundaries. States will control state waters and the effect of development on this resource. States may also have lists of specific endangered species and endangered plants that need to be considered during the site analysis review. These sensitive environmental issues can be used as a benefit to the project. The environmental attributes of the site can be positive marketing aspects.

Most states approve water and sewer public facilities. States allocate utility permits for water and sewer facilities by using existing infrastructure to guide community growth. Areas outside of public service areas must use alternative methods of providing water and sewer facilities. These alternative methods, such as septic systems, can also be managed by state agencies. Individual wells providing potable water to each dwelling are permitted by state agencies. States monitor and manage the drawdown of aquifers caused by multiple individual wells or by expansion of public facilities serving new housing developments.

State-mandated building codes based on the model building codes are implemented through state agencies but controlled and managed by local building departments. These code provisions also affect the way subdivisions are designed.

Federal Government

Several federal agencies—the Department of Housing and Urban Development, Department of Energy, and EPA—directly affect housing projects and the use of land for developments. Federal agencies are responsible for the protection of endangered species and endangered plants listed on the federal registry. The US Army Corps of Engineers (USACE) has regulated wetlands and floodplain areas for many years. It is clear that the permits needed to alter a floodplain for development purposes may become the critical path in terms of time. The USACE requires a Conditional Letter of Map Revision and a Letter of Map Revision for any impact of a floodplain. It is important for the engineer to communicate the ramifications of seeking these permits.

Some states have taken over the jurisdiction of the implementation of wetland protection. Since the passage of the Clean Water Act, the regulation of stormwater, wetland, and environmental issues for subdivisions has been elevated to a much higher plateau of sensitivity. Another federal agency having an effect on development operations is the Occupational Safety and Health Administration and the Federal Aviation Authority.

Subdivision Approval Process

A typical subdivision approval process, from inception to final approval, is presented to illustrate the various levels through which any housing project most likely must pass. This description does not clearly indicate the amount of work and effort that private industry must accomplish over and above what is shown in Appendix 3-5 just to obtain market acceptance.

Before the subdivision approval process starts, a builder/developer must invest a tremendous amount of effort in investigative work, contract negotiations, and value judgments to allocate financial resources for the submission of improvement plans. Even after a builder/developer completes the regulatory process and obtains governmental approvals, the marketability, success, or failure of a project will not be known until model homes are constructed and for sale. The public ultimately judges the marketability of the product because people will look and buy—or just look!

Typically, 18 to 30 months is the timeframe from finding land to giving the keys to the first homeowner. In some states, reaching the point of closing a home takes less time and fewer financial resources to complete a project than the national average. However, in many states, the regulatory process continues to be more complex and more layered, so the development industry is viewed as a high-risk industry in which success is earned. The philosophy of developing small parcels for a relatively small number of homes, even up to 50 houses or 50 lots, provides an excellent inventory of approved and unimproved land. The financial resources to build out and to sell out a subdivision of up to 50 lots can be managed by many small- to medium-sized development firms or building companies.

Government reviews and approvals and the issuance of permits are imbedded in the bureaucracy to control the development pattern of a community. In fact, many processes continue to evolve as special interest groups lobby government to influence the development process. The success of the industry rests on the shoulders of those who understand and seek solutions within the framework of government regulations.

Planning Board Guide for Citizen Participation

The regulatory climate and atmosphere are driven by four elements:

1. Planning board members
2. Municipal professional staff members
3. Applicants and their design team members
4. Members of the public.

Most planning boards encourage and welcome citizen input at planning board hearings. In fact, public comment sessions, including question-and-answer periods for the general audience, are commonplace. Unfortunately, most municipal governments do not educate the citizenry about proper hearing procedures and the laws that govern planning board operations and management responsibilities.

An engineering firm may consider preparing a citizen's guide to planning board hearings that is geared to the specific community. By making such a document available to the public, interested citizens can read and understand the hearing process. Such a document could include the sections that follow.

Application Process

Define some of the major elements of a housing, commercial, or industrial project that are reviewed and approved by planning boards. Include such definitions as *site plans, minor* and *major subdivisions, master plans,* or *concept plans,* any of which might be used on a planning board agenda. A description of each application type would be appropriate, and also include other governmental agencies required to review each type of application.

Associated Agencies

Include the role each agency has in the subdivision approval process. Provide contact names and phone numbers for all of the agencies involved, including governmental departments, such as the engineering department, fire safety department, traffic safety department, planning department, environmental commissions, shade tree commissions, water and sewer districts, soil conservation districts, and other utility companies—gas, electric, phone, and cable.

Procedures

Outline the exact procedure in which the municipal government reviews subdivision applications. As an example, the applicant submits a certain number of plans, applications, and fees associated with a subdivision plan. The package is submitted to the planning board secretary 10 to 14 days before the next regularly scheduled planning board meeting. The application is reviewed for completeness to ensure that all the information is contained in the submission. Once the application has been deemed complete, an explanation of the distribution of the plans should be outlined in detail. The time permitted for the review by each agency should be indicated. Include the statement that all plans, reports, environmental impact studies, traffic studies, and drainage reports are made available at the planning board office during regular business hours for review by the general public. After the application has been deemed complete and reviewed by the governmental staff, the application is scheduled for a public hearing.

Types of Approvals

Describe the types of approvals rendered by the planning board, such as preliminary approval, final plat approval, or approval conditions. Include on what days the planning board meets, what time the meeting starts, and what time the meeting is adjourned. Indicate how the agenda is arranged, such as first-come first-served, or if the application has been delayed, postponed, or moved from one month to another because of legal notice problems, and so on.

Types of Meetings

Define the different types of meetings. Many planning boards hold work sessions, informal hearings, and formal hearings. A flowchart illustrating the application process and the type of hearings would be helpful and easily understood by the public.

Description of a Hearing

Explain what the public will witness at the public hearing. As an example, explain that a planning board secretary keeps minutes, handwritten or audio recorded. There will be planning board officials, a planning board attorney, and professional staff members providing technical input and review documents. Also attending will be the applicants, design team members, attorneys, and the public. Most planning board meetings will be live streamed on the city's or town's website.

Traditionally, meetings start with the Pledge of Allegiance, a reading of the previous meeting's minutes, an approval of invoices, the memorializing of previous resolutions, and the addressing of any issues or topics of interest, which may include master plan issues or ordinances. An explanation of each component of the agenda will help the audience understand the workings of a planning board hearing.

Presentations

Include a section on the applicant's presentation about how the board interacts with the applicant and the design team members or expert witnesses. Clearly indicate for the public what to expect during the presentation and review of a specific application. The applicant will most likely be the builder or developer; the attorney or engineer will present the project specifics. Applicants will try to present the project in a clear and professional manner. The engineer will most likely present the technical aspects of the project. The application must be presented methodically to illustrate how the project benefits the community as a whole. The intent of the project is outlined in detail, and what the public can expect once the project has been completed will be described.

Once the presentation of the applicant is completed, the planning board members and agency professionals have an opportunity to ask specific questions about the project design. The applicant will respond to each question in sufficient detail. If the question is phrased to allow a yes or a no answer, the applicant will usually respond in that fashion. The chair of the planning board will keep the hearing running efficiently and will strive to keep the discussions on point, brief, and detailed as much as possible.

Questions and Answers

Explain that once the board members and staff have completed their reviews and discussions, the chair will open the public session for questions and answers. Include a list of dos and don'ts for the public to observe during this portion of the hearing. In general, the public forum will last as long as there are citizens willing to ask nonrepetitive questions. The public session is for asking questions and not for making speeches. The chair should not permit duplicate questions by the public. The public should maintain a sense of decorum regardless of the emotional aspects of the project. The chair must keep control of the meeting and not permit unruly participation by either the applicant or the public.

Work Sessions

Describe how successful municipalities have work sessions or workshops with the applicants to discuss a project design with several members of the planning board and staff. The forum is usually an informal discussion, and the opinions and discussions are nonbinding for the applicant.

Definitions

Expand the brochure to include definitions of terms usually heard at planning board meetings (e.g., waivers, zoning ordinances, bulk requirements, resolutions, design exceptions, variances, and engineering terminology). Include the application forms and fee schedule that applicants must complete before the project is accepted for review by the municipality.

A planning board public awareness document is welcomed by individuals in the arena. When the public is informed of the process, they become part of the process and will respect the system and the efforts of all parties involved. When all parties, including the public, adhere to the rules of the planning board, the meetings and presentations will proceed efficiently and without undue interruptions. After all, time is money for the applicant, and any delay affects the success of the project.

Chapter 17

Municipal Government's Changing Landscape

Stress and demands brought on by the need to have a municipal government that meets the needs of the existing citizens and plans for growth is monumental. Suburban municipal governments have managed the growth much better than communities located closer to an urban core—not so much anymore. Those municipalities with solid foundation of infrastructure and within a reasonable commute to the urban core are now being noticed by the real estate industry. It is common for major national builders to expand into surrounding passed-over communities to meet the growing demand for housing—for both new homes and apartments.

The national landscape is changing rapidly, from Amazon affecting the retail industry to work-from-home options made available by employers. Municipal governments are not structured to quickly manage the changing landscape. Local governments are in a no-win situation, because government is not proactive but reactive. Local communities can be categorized by the following four descriptions:

- No-growth communities are typically pro-non-residential and anti-housing because they want to establish a tax base and they consider housing to be a drain on the general fund. This is valid to a certain extent but may be shortsighted if the emphasis is on retail and commercial. A community may lose its sense of identity. Rezonings in a no-growth community are no longer practical in terms of time and resources. There are residential projects that pay for themselves.
- Slow-growth communities are typically overregulated, and it takes all too long to obtain permits and approvals. It has little to do with the market or lack thereof, but an overregulated real estate industry will surely pass over a marginal market with lower land prices in a slow-growth community for better located property, regulatory

certainty, and market acceptance. Slow-growth communities will eventually sway either to no-growth or pro-growth.

- Pro-development communities are becoming more apparent in the changing real estate landscape. These communities work well with the land development industry to achieve projects boosting the local economy while meeting market demands. Many pro-development municipalities are those communities that were previously passed over but have an excellent balance between residential and nonresidential uses. Because of the rapport between public and private interests, the community thrives and reaps the benefits of a proactive community vision, even when many of these communities are challenged by annexation requests.
- Smart-growth communities have embraced development trends that are marketed as sustainable. Many of these jurisdictions embrace best management practices and cluster zoning as their definition of sustainability. The flexibility of land uses to meet market demands and the desire for creative design are forgotten in smart-growth communities (Figure 17-1).

Local governments prepare a variety of overall community plans to guide their community into the future. Many communities unfortunately do not update these plans on a defined timetable. It is important to update the plans and/or ordinances as the municipal landscape changes, but many communities will not allocate the funds necessary to do so. Master plans, vision plans, zoning maps, comprehensive plans, transportation plans, subdivision ordinances, and land development guidelines are all used in controlling growth patterns.

Architectural Guidelines and Price Points

Starting in the mid-1990s, local governments changed their approach to real estate development, both residential and nonresidential. In part, government's response was a reaction to development pressures. The public also has been established as a key vocal stakeholder in the regulatory process. The trend at planning or zoning boards started innocently by asking the builder/developers the price range of the homes to be built or whether any of the retail or office space had already been leased, they asked about the square footage of the homes or about the HOA documents and the controls that the builder/developer imposed on the aesthetic appearance of the project.

Figure 17-1. Local government entities will always control and guide residential developments because housing is a micro-market phenomenon. Discovering the philosophy of the politicians toward housing development will assist in determining the practicality of the unit count and project schedule.
Source: David E. Johnson.

Architectural or aesthetic zoning was born in the early 2000s when planning and zoning boards started to listen to stakeholder views and opinions, which were primarily based on emotion rather than on fact. As a result, local government requested builders to change products, materials, price points, and so forth. The ultimate goal was to seek improved appearances of the homes and increase their value. In return, homes with higher values paid higher real estate taxes.

Because builders could not and cannot afford a delay during the regulatory process, they agreed to local governments' architectural and aesthetic requests. Builders changed their housing product specifications (e.g., granite countertops as standard), materials (e.g., brick as standard), and product designs (i.e., larger houses on smaller lots). Government asked, requested, or demanded, and builders agreed.

Municipalities are now moving toward using a form-based code rather than the traditional zoning code. A form-based code is more suitable for urban settings, but the intent of complementing building forms with outdoor spaces and infrastructure may eventually be the basis for residential developments. Large-scale residential communities could benefit from this type of an approach, as standards are set specifically for large community-style projects.

Residential Roads and Traffic Impact Studies

Local governments also started to change infrastructure standards, such as increasing roadway widths, wider sidewalks, more restrictive utility design standards, expanding water quality and quantity control measures, and so on: these improvements were shouldered by the building industry knowing local government's lack of the funds for maintenance. Regardless of the number of units, builders are required to submit a traffic impact analysis (TIA) prepared by a traffic engineer retained by the builder or by local government. The conclusions reached by the traffic engineer dictate the off-site road and intersection improvements to be funded by the builder.

Local government will not invest in road or intersection improvements for a yet-to-be determined impact. Many local governments have a road assessment report outlining the pavement condition for each public-owned street. This report is the basis for the capital road maintenance plan. As more residential projects are accepted, newer roads built with new standards will have a longer life and require less maintenance.

Off-site road improvements may require right-of-way acquisitions, rerouting utilities, or replacing existing pavement. It is best to negotiate a developer's agreement outlining a payment schedule for funding the improvements. By linking the payments to closings, builders will better manage cash flow and government would be responsible for the improvements. Typical road improvements would consist of turn lanes, traffic signals, and road widening along the property frontage. In some situations, a new primary road shown as a line on a transportation map may be required. Cost sharing and timing for this level of improvement is critical to the overall success of the project.

The national fire code overrides local regulation in many instances because councils approve the use of the code without fully knowing the ramifications of the provisions. Fire departments are implementing regulations that may be inconsistent with local requirements. As an example, fire departments are requesting wider road widths that are inconsistent with sustainable design, slower vehicle speeds, and pedestrian friendly designs. Fire departments are also requesting wider side yard setbacks to further separate structures and limiting the number of homes with only one ingress/egress. The design engineer must also be cognizant of the fire code provisions and the implementation of those provisions at the local level.

Municipal Services and Cost Benefit Analysis

Projects located outside of the city limits require annexation agreements and a cost benefit analysis. The cost benefit analysis is typically prepared by government to determine the fiscal impact the project will have on government's operations. The higher sales price of the homes will result in higher real estate taxes and increase the municipal general fund. Higher revenues allow local governments to better keep pace with housing project buildouts.

Administration, police, fire, and parks departments are rarely affected by a new housing project. However, public works is most affected by any real estate project. The cost of providing water, sewer, and solid waste pickup is weighed against the revenue generated from these services. In most cases, new residential projects break even with water and sewer services but not with solid waste pickup. The approval of a residential project either rests with the city council or planning board, and the cost benefit analysis is important to know, but if the project is "good" for the community, the project will likely be approved.

Economic Development

There has always been competition among states, counties, and local government to attract companies for their jobs and tax revenue. In many cases, economic development groups are quasi-government entities having the support of governing bodies encouraging companies to expand their operations or relocate facilities to new locations. The competition for these companies will become more intense as the economy expands.

Over the last decade, Texas added more jobs than the rest of the states combined owing to a pro-business environment. In states or local municipalities not structured as pro-business, governments will resort to making offers companies cannot refuse. In some cases, municipalities will use eminent domain to amass land for large business complexes. Some local governments will use tax or other incentives to make a deal. In any case, *land* is the important component, that is, location, existing infrastructure, and so forth. As an example, the Charleston County Aviation Authority *leased* 240 contiguous acres to The Boeing Company. Not having to purchase the land for development purposes is significant. Boeing received other incentives to locate its plant in Charleston, South Carolina. Local municipalities must determine the unintended consequences economic development will have on existing infrastructure and municipal operations.

Stormwater Facilities Operation and Maintenance and MS4 Permits

The Federal Water Pollution Control Act of 1948 was the first major US law to address water pollution. Growing public awareness and concern for controlling water pollution led to sweeping amendments in 1972. As amended, the law became commonly known as the Clean Water Act (CWA).

The 1972 amendments

- Established the basic structure for regulating pollutants discharges into the waters of the United States,
- Gave EPA the authority to implement pollution control programs, such as setting wastewater standards for industry,
- Maintained existing requirements to set water quality standards for all contaminants in surface waters,
- Made unlawful the action of any person to discharge any pollutant from a point source into navigable waters, unless a permit was obtained under its provisions,
- Funded the construction of sewage treatment plants under the construction grants program, and
- Recognized the need for planning to address the critical problems posed by non-point-source pollution.

EPA is strengthening the stormwater program and finding ways to support local communities to manage their stormwater programs. EPA has implemented the Municipal Separate Storm Sewer System (MS4) program as mandated under the CWA and initiated in 1990. MS4 is required for all operators located in areas designated as urban. There are two tiers to this program: Phase 1 for populations of more than 100,000, and Phase 2 applicable for smaller municipalities. The purpose is to manage stormwater to minimize or reduce pollutants in stormwater runoff before the flow reaches federal waters, protect water supplies, and comply with the water quality standards established by EPA. This program has been delegated to states for implementation and monitoring. The program will only receive more and more attention as urbanization continues to expand into suburban areas. Most MS4 programs are funded by a fee charged to property owners based on the amount of impervious surface. Many of the smaller municipalities required to comply with the MS4 requirements have a difficult time raising the funds necessary to properly manage the program at the local level. There are now regional groups of municipalities contributing expertise and developing regional approaches on which all the municipalities can rely to comply with the requirements.

The MS4 program is based on the implementation of six minimum control measures as stated in EPA Fact Sheet EPA 833-F-00-002. These measures are outlined as follows:

1. Public Outreach and Education
   Distributing educational materials and performing outreach to inform citizens about the impacts polluted stormwater runoff discharges can have on water quality.
   This control measure is for both residential and nonresidential projects. As residential lot sizes continue to shrink, the amount of impervious surface still is less than most nonresidential projects. Essentially, municipalities must find the appropriate means conveying the requirements of the program to the entire community. Appropriate means consist of print media, signage, websites, advertisements, interviews, events, and so on for the distribution of information necessary for citizens to be aware of how their actions may contribute to the degradation of the quality of water in waterways.
2. Public Participation and Involvement
   Providing opportunities for citizens to participate in program development and implementation, including effectively publicizing public hearings and encouraging citizen representation on city committees, citizen advisory groups, or boards.

Citizen participation would increase understanding of the program and why it is being implemented. This strengthens the public support of the program. Not only is this partnering valuable for all citizens, but it provides a forum for other entities to become involved in the program outcomes.

3. Illicit Discharge Detection and Elimination

   Developing and implementing a plan to detect and eliminate illicit discharges to the stormwater system.

   As defined by the program, an illicit discharge into a storm drainage system is a discharge into the system that is not entirely of stormwater runoff. The municipality or permittee is also responsible for implementing a proactive illicit discharge detection elimination (IDDE) program. IDDE is meant to stop activities that may result in an illegal discharge, that is, an illegal connection or dumping into the storm drainage system. Certainly, public education and awareness will over time minimize illicit discharges, but completely stopping illicit discharges is a failed expectation of the program.

4. Construction

   Developing, implementing, and enforcing an erosion and sediment control program for construction activities that disturb more than one acre of land.

   This element of the MS4 permit has been in practice for decades. Granted, in some areas the attention to controlling the runoff from construction sites has varied considerably. Some areas pay more attention to the design and permitting but lack the resources to inspect and monitor the measures during construction. The debate has always been the level of protective measures imposed against residential projects versus those measures required of farming activities. The debate will continue, but for residential projects, the primary sediment and erosion control design elements consist of temporary seeding practices, tacking pads at construction entrances, phasing of projects limiting disturbed areas, silt fences, diversion ditches, and sedimentation ponds.

5. Post-Construction Runoff Control

   Developing, implementing, and enforcing a program to address discharges of post-construction stormwater runoff from new developments and redevelopment areas. Structural quality control measurements, ordinances, and protecting sensitive land forms are applicable controls.

   Stormwater management facilities managing the quantity of discharge from the site to mirror predevelopment flows has been in practice since the 1970s. Municipalities have expanded their focus from *quantity* of runoff to the *quality* of the runoff leaving a construction site. In some states, the quality of runoff has become much more important than the increased flow, because impervious surface is being limited in residential projects. In fact, some municipalities track the amount of impervious surface by lot and in total. This approach will eventually become standard practice only if public monies and staff are available to make it happen on a consistent basis.

Ownership and maintenance of the entire storm drainage system remains problematic for municipal governments and homeowners. Nonjurisdictional streams and drainage systems beyond the public rights-of-way eventually require storm drainage to be considered an essential utility to be owned and maintained by local government. Storm drainage systems, stormwater management systems, and best management practice (BMP) facilities should be owned and maintained by local government.

6. Good Housekeeping and Pollution Prevention
   Developing and implementing a program with the goal of preventing or reducing pollutant runoff from municipal operations. The program would include staff training on pollution prevention measures.
   Equipment storage areas; maintenance garages; landscape material lay down areas; chemical buildings; and materials used for road, water, sewer, and drainage maintenance yards shall be designed and maintained to prevent the discharge of pollutants from those areas into a receiving waterway. Local governments must be a leader in good housekeeping measures to illustrate the importance of the MS4 program.

The regulatory stormwater initiative over the last 40 years has made a difference in the quality of the nation's waterways. There are noteworthy problems that still remain for local government to address: operation, maintenance, and ownership. For nonresidential projects, the owner of the project owns, operates, and maintains stormwater control facilities.

However, for residential projects, local government does not have the resources to operate and maintain these facilities. Local government rarely takes on the responsibility of owning, operating, or maintaining this utility due to funding, liability, and staffing issues. In fact, resource allocation to manage stormwater as a utility is not significant enough to monitor the facilities on a regular basis. It is clear that the lack of attention and proper maintenance of these stormwater facilities results in a system failure, which ultimately defeats the purpose of the original intent.

Those municipalities that have MS4 permits must find the financial means to elevate stormwater to public works utility. As an example, in North Carolina, MS4 permit holders charge a monthly stormwater fee that is added on the monthly water bill. The revenue generated from this fee funds the stormwater department responsible for this utility. Most stormwater departments are reactionary to visible failures from minor flooding, sink holes, and erosion.

Storm drainage systems are costly, taken for granted by the public, and not seen unless a failure occurs. Thus, ownership philosophy should change so that government accepts storm drainage systems as a public utility. Once the overland flow reaches a defined conveyance system within a public right-of-way or public easement, local government accepts the ownership and maintenance of the storm drainage systems to a positive outlet. Drainage systems would include swales, ditches, curbs and gutters, roads, inlets, manholes, pipes, open channels, and culverts. A positive outlet is a jurisdictional stream or waterway.

Water Distribution and Water Supply

Municipal governments provide a service of delivering potable water to their customers. The public water system supplies water at a pressure, rate, and quality that satisfies customers at a reasonable cost. Municipal governments allow fire departments to use the water system when needed. Water is a protected resource and a life necessity demanding exceptional management of its use. Builders take water supply and distribution for granted as part of the project. If a larger water line or water line extension is required to serve a new residential project, builders will install the necessary water system at their cost and donate the lines to local government without much debate.

Wastewater Treatment Plant and Line Capacity

It is rare for any state to build a new wastewater treatment plant. A recent exception was in Watkins Glen, New York, opened in 2020. Many wastewater treatment plants are nearing the end of their useful life, and others are reaching their maximum capacity. With rivers and streams being assigned new pollutant loads and limits, the ability for a local government to expand its existing plant or build a new facility is very difficult because of cost, permits, and time. Communities will grow either by annexations or through redevelopment. One million gallons per day of plant capacity equates to about 4,000 housing units. Rezoning generates flows from different uses, and local governments must understand the long-term consequences of rezoning actions that may increase the flow and decrease the plant capacity available for subsequent projects. Although plant capacity is a long-term and on-going issue, due-diligence efforts must also analyze the collection system from the PIQ to the wastewater treatment plant.

The plant may have capacity but the trunk lines to the plant may not be of sufficient size to convey the flow from a new residential development. It is important to know the pipe sizes and available capacities during the due diligence period. However, a fair share contribution based on new flow is reasonable to ensure new parallel sewer lines or new trunk lines are available to serve the project.

Chapter 18

Land Development Bid Process

The bid process is the bridge between the regulatory approval process and the construction of the project. This chapter discusses in depth a step-by-step procedure for managing the bid process so you can know your costs, schedule the work properly, understand contractor capabilities, and construct a quality subdivision. This residential land development methodology can give your project the highest chance of staying within budget and adhering to the project schedule. Using current project management software will provide the tools necessary to manage the information flow.

Plans and Specifications

The construction documents, as amended during the final design phase of the regulatory permit process, become the approved final construction documents that you will use to build your subdivision. These construction documents, once approved, should be stamped for construction, dated as of the last revision date, and signed and sealed by the professional engineer in charge of the project. Owners should accept only design plans stamped "For Construction" for distribution to contractors. This documentation provides the contractors with the knowledge and comfort that the plans have received local governmental approvals.

Municipalities may require signature blocks on the construction documents for the planning board chairman, secretary, and municipal engineer. They sign the original plans, indicating their approval of the construction documents. Even if a locale does not have this mechanism in place, it is a good idea to provide a signature block for future reference. This reference is especially important when the municipality accepts ownership and maintenance of the project.

Residential subdivisions should have their own land development specifications, as shown in Appendix 18-1. This outline indicates the various headings incorporated in the specifications report. Different from specifications that relate to interstate highways or major public works programs, these specifications are consistent with the activities required to construct residential projects. The specifications are owner friendly and put most of the construction and design responsibilities and liabilities onto the appropriate parties. These specifications, if implemented in the field and monitored by the owner and the local municipality, provide a framework for building a quality subdivision.

Prebid Conference with Staff

Before sending the plans and specifications out for land development bid proposals, schedule a prebid conference with the operational staff responsible for overseeing the construction. A brief outline of the prebid conference is provided in Appendix 18-2. Discuss in detail the project specifics and construction requirements. Begin with clearing operations and follow through to the construction sequence of events. During this prebid conference, any problems associated with the construction documents and the adherence to project schedules should be identified and discussed in depth.

If the project is to be phased, the location for the storage of materials, the staging areas for equipment, and the delivery of the first housing construction pad should be discussed. Land development activities for a residential project should complement the housing activities of the builder. Discuss a definite timetable and project schedule to determine if any unusual hurdles or extenuating circumstances may affect the schedule and the delivery of the first housing pad. The corporate objective is to deliver pads on schedule to sell lots or construct homes.

Ask staff members for recommendations on which contractors should be invited to bid on the project. Discuss the benefits of hiring separate contractors with specific expertise (such as utility installations, curbing, or paving) versus hiring a contractor responsible for the entire project who subcontracts out for specialty items. Hiring separate contractors for various activities can provide a lower overall project cost. However, this has to be weighed against the need for additional on-site supervision. Moreover, any failure of infrastructure will be subject to debate as to the responsible party when multiple contractors are involved.

If the contractor retained to build the project subcontracts out the work, the owner should hold the contractor responsible for any failure to the infrastructure. In considering available contractors, price is certainly a consideration, but the delivery of pads for housing construction generates the cash flow. The first-phase housing lots should be constructed on an expedited schedule. The design plans must be sensitive to the phasing objectives of the company. Once the first-phase pads have been constructed, the land development infrastructure must be completed in a way that eliminates conflict with housing subcontractors. This allows closing of lots while subsequent phases are being constructed.

Bid Process

Contact the contractors selected to bid on the project by phone or letter and ask if they would like to bid on certain activities or on the entire land development work. It is appropriate to charge a fee for the project plans given to contractors. This fee defrays the cost of printing the construction documents. The fee could be considered a deposit if the plans are to be returned. Returned plans can always be used during construction.

Contractor Information

Each contractor should complete the contractor information sheet, as shown in Appendix 18-3, and return the form with the bid proposal. This information sheet collects data on the financial capability of the firm and a list of references. These references should be contacted. Always consider new firms so as to expand the bid list. Most new firms are motivated to do work on time and within budget. For new companies, check references and visually inspect their previous work. Also, seek a personal commitment from the owners of the company to meet schedules and price agreements.

Contractor Meeting

Before the bids are submitted by the contractors, schedule a meeting with them to present the project goals and objectives. This meeting involves all contractors asked to bid on the project, giving them the opportunity to see and hear their competition. The contractors should have the construction documents at least one week before this general meeting, so they have time to review, analyze, inspect, and understand the intent of the project. At the contractor meeting, discuss project details, scheduling, phasing, and design elements. In addition to the contractors, invite local utility company representatives, municipal inspectors, the design engineer, and the project architect to attend. This way, everyone has the opportunity to openly discuss the design specifics of the project and the construction requirements for delivering the first housing pad. Allow sufficient time for all contractors to ask detailed questions.

This contractor meeting gives each contractor sufficient time to fully comprehend the project parameters. Thus, during the analysis of the bids, discussions on any unknown or unusual factors should already have been addressed. The importance of having the competitors ask questions and face off with their rivals is to ensure that potential problems are addressed before they become actual problems. Also, knowing their competition, contractors will provide a competitive price before entering into the negotiation phase.

During the contractor meeting, establish clear timetables for the submission of bids. A definite deadline for the submission of bids is appropriate, and it should be stressed that any bid submitted after that date will not be accepted. The contractors should be involved in setting the deadline. Thus, if a contractor does not meet the deadline, it can fairly be interpreted as an indication of their inability to meet schedules and perhaps their inability to perform the land development activities. Schedules in land development and housing projects are key to carrying out any successful project and sustaining positive project cash flow. Builder/developers rely on contractors to perform efficiently, on time, within budget, and with a high degree of quality.

Contractor's Schedule

A builder/developer's reputation is established or tarnished by contractors implementing construction documents in the field. The contractor should submit the project schedule with the bid proposal. This schedule outlines the elements of the work in a timeframe that can be analyzed against corporate goals for delivering houses for closing. The schedule is one aspect of a review of the contractor's manpower, quality of construction, reputation, and ability to perform in the field. Their schedule becomes their personal commitment to complete the work on time (Figure 18-1).

Figure 18-1. Project schedule. Success in land development is associated with a defined schedule implemented by the contractor. Selecting a contractor is based on cost and the commitment to complete the project on schedule.
Source: David E. Johnson.

Bid Forms

Consider preparing bid forms for the contractors to use in submitting their prices. This bid form shows the various work activities. Either leave the unit quantities blank or provide the quantities in the bid form. In either case, the design engineer should be providing a detailed quantity report for the project using the same activities required of the land development contractors. The design engineer could provide unit pricing but it is not necessary because the pricing is to be provided by the contractors for their specific work responsibilities.

By providing quantities in the bid form, those quantities supplied by each contractor can be compared in detail. Thus, it is easier to analyze the individual bids to obtain a true comparison of costs. If the quantities are not included on the bid form, land development contractors will organize their bids by altering quantities and unit pricing to suit their operational style; they will also indicate any problems with the quantities. Leaving quantities off the bid form makes it very difficult to properly compare bids. The owner or builder/developer will be required to engage in substantial discussions with all bidders to ensure the comparison of each proposal is completed with fairness.

Bid Analysis

On receipt of the bid proposals, use an accounting sheet or a computer spreadsheet to enter in the quantities, unit prices, and total prices for each element of work. If the quantities were provided on the bid form, the analysis is reduced to reviewing the unit pricing for each category; if the contractors do not alter the quantities, you can conclude that the quantities are accurate. If the quantities were not provided on the bid form, it may be difficult to compare the work activities and prices among contractors. In most cases, the engineering company should be able to provide accurate cut-and-fill quantities for the project design. Most earthwork contractors use software to provide a computerized printout of the cut-and-fill analysis for the project. Review this information against the quantities prepared by the engineers. If there are no discrepancies or debates over the quantities, your analysis will center on contractor pricing, scheduling, available workforce, and experience.

Prepare a spreadsheet with the contractor unit prices and extend the quantities to prepare a cost estimate for each land development activity (Appendix 18-4). The bid analysis can be entered into the computer or simply done by hand. In reviewing each bid proposal, resolve any bid discrepancies to ensure a true comparison among bid proposals. Any bid discrepancy, if significant, should be presented to all companies so that there are no underlying problems during the analytical part of the project. Because this is a private contractual arrangement, public bidding laws do not regulate the process. However, it is good business practice to ensure the bid comparisons are true comparisons; it is better to make decisions on a monetary basis and not on a personal basis.

After spreading the schedule of values, double-check the unit prices, extensions, and any notations the contractor made to ensure that the contractor understands the scope of the work. If misunderstanding of the work scope seems to come to light in the bid proposal, contact the contractor to resolve the issue.

Once the schedule of values has been extended and all discrepancies resolved, analyze each bid proposal and prepare a detailed recommendation on the contractor selection. Prepare negotiation strategies for each contractor selected to do the work.

Contractor Selection Process

The following are key elements of the contractor selection process: competitive price, sense of cooperation, skill, expertise, project experience, reliability, references, professionalism of the owner and experienced management staff, quality equipment, and qualified superintendents (Figure 18-2).

The contractor must have a qualified superintendent on the jobsite, because the builder/developer will be working with that person on a day-to-day basis. The superintendent must be sensitive to the owner's needs during construction. If the contractor's superintendent has been delegated the responsibility to build the project and his or her decisions

Figure 18-2. Contractor qualifications. The land development bid process includes an investigation of the contractor's capabilities, experience, and reputation for quality construction of residential neighborhoods.
Source: David E. Johnson.

are questionable, the contractor should be asked to assign someone else to manage the project.

If several contractors are within a reasonable range of total dollars, those companies should be asked to proceed with the negotiation of a contract price. The intent is not to produce a bidding war, but it is incumbent on the builder/developer to obtain the best price for the work. Some contractors may view the project as important to meet their corporate goals; they may be more willing to work with less of a profit margin simply to get the work. Land developers need to keep their equipment and their workforces working continuously. The automatic selection of the low bidder may not lead to the best result. It is appropriate to call two or three contractors to further negotiate the final contract agreement.

If separate contractors are selected for different land development activities, such as for curb or paving work, limit the negotiations to the specific work items. The combined individual contractor prices should be compared to the general contractor's price to do all the work under its corporate umbrella. This approach provides a true comparison of the price for a group of individual contractors versus the price from one contractor. The cost difference must be weighed against the management responsibility required to oversee multiple subcontractors.

If the bid proposal includes the owner's quantities, the unit prices for each activity require negotiation. Sometimes it is cost effective to negotiate a lower price on a specific line item having a large quantity. The cost savings may be much greater than what could be achieved by negotiating an unreasonable unit price for a small amount of work.

During the negotiation phase, meet with the contractors to discuss their bid proposals and obtain the personal commitment from each contractor to adhere to the project schedule. This personal commitment is important to the overall success of the project. A well-constructed project is hardly ever recognized by its quality, but it is judged by the number of complaints. Eliminating problems with strong management will be the only reward a contractor will receive. If the project is critical to business objectives, provide a timetable with financial incentives to the contractor. If quality construction can be accelerated ahead of schedule, consider offering a financial award to the contractor.

Land Development Contracts

The preparation of a land development contract for execution by the owner and land development contractor can take several forms.

Unit-Price Contract

A unit-price contract establishes specific unit prices for specific work scopes, and the quantities of material installed are measured in the field. The agreed-on unit prices are used to calculate the ultimate payment for work accomplished.

Time-and-Material Contract

A time-and-material contract is based on labor, equipment, and material. The contractor provides an equipment list and the hourly cost for equipment and labor (e.g., the equipment operator). The equipment time and material used become the basis of the contract. In general, in a time-and-material contract, the equipment list and the hourly rates combined with receipts from material suppliers represent the total cost for the activity. The time allocated by a contractor should be verified by time cards or staff monitoring. The contractor should not pass the material costs through their administrative procedures and add a fee for processing. The builder/developer should consider opening an account with material suppliers (pipe suppliers, concrete companies, and others). A true savings can be achieved by purchasing material directly from the suppliers. Managing the material used by a contractor will result in cost savings.

Lump Sum Not to Exceed

A "lump sum not to exceed" contract establishes a fixed total amount. If properly administered, this type of contract can provide the highest degree of predictability of the project costs. The "lump sum not to exceed" contract is extremely beneficial to the developer in controlling costs, knowing costs, and managing costs for the project, because the actual profit margins accruing to the developer can be anticipated in the project pro forma. Extenuating circumstances in the field under a lump sum contract may become the source of debate between the developer and contractor for payment of extra work. However, the possibility of extenuating circumstances can be overcome by including the equipment list and the hourly rates as part of the lump sum contract. Another potential problem with a lump sum contract comes from the complexity of earthwork operations. Unsuitable soil or rock encountered during construction and not identified in the soil investigation work may cause a problem for the contractor. Therefore, during the negotiation process, address this situation as an extra work provision in the contract. The contract provisions should be specific in terms of extra work activities; then, as long as the extra work is approved by the developer, the contractor can proceed without delay by using the agreed-on hourly rates for equipment and operators to complete a task.

For budgeting purposes, extras can be handled several ways. A contingency budget item can be included in the overall project pro forma, which has its own separate account. An alternative is that each individual land development activity can be budgeted with its own contingency line item. The total land development cost, known or projected with accuracy, is a must before the release of houses or lots for sale.

Once houses or lots are released for sale, the project unknowns, resulting in extras, or the inability to perform the work in an efficient manner will increase the development costs. If these costs are not accounted for in the pro forma, the price of lots or houses must be increased. It is extremely important to maintain continuous building operations so that profitable companies can continue to prosper and provide new housing opportunities.

A comprehensive 19-step contract outline highlighting the provisions of a land development contract is provided in Appendix 18-5. The contract should clearly identify project schedules as being an important element of the work requirements of the contractor. A sample land development contract, as shown in Appendix 18-6, has proven to be successful in implementing a single-family residential project. Once the contract has been executed, schedule a preconstruction meeting. At the meeting, include the staff, land development contractor, surveyor, utility companies, engineer, governmental officials, and others to discuss the project specifics, mobilization of the land development contractor, and the staging area within the project site. Any last-minute contractual problems can be resolved at that time. Subsequently, regular project meetings are necessary to ensure open communications among the contractor, government inspectors, and developer. Addressing problem areas as they are presented in the field is necessary if the project is to maintain its schedule.

External influences will affect the project schedule. However, these situations, such as inclement weather, cannot be controlled by the owner.

As with any project, the schedule should include slack time that accommodates lost days resulting from uncontrollable and unpredictable events. Adverse weather can alter the schedule but should not change the goal of delivering the first housing pad on time.

Construction Document Revisions

The construction documents for any residential housing project may need to change for a variety of reasons. Changing the construction documents requires the engineering company to assess the change and revise the plans. Revision dates, revision numbers, and reasons for the change should be indicated on the plans. The revised plans should be sent to the appropriate governmental agencies and to the land development contractor for implementation.

Field Conditions

Field conditions affecting the project could cause revisions to the construction documents. For example, if a topography bust is discovered and the elevations in the field are not consistent with the design plans, the construction documents must be reviewed and, if necessary, changed to reflect the existing elevations. The land development contractor will benefit from this field problem. Do not delay the project while negotiating the cost ramifications for correcting a field problem.

Better Designs

Owner representatives may change the scope of the work or change the construction documents to incorporate a housing design or foundation design that better reflects the project goals. The housing product should change as market conditions change. If a developer concludes that the project should be modified to accommodate a different product approach for greater absorption, the developer can issue a design change notice. For example, preparing pads for slab construction would alter earthwork quantities if the plans originally called for basement foundations and the negotiated price was based on basements.

Contractors should submit operational design changes through a shop-drawing procedure. If they find a less expensive material or a better way of completing the work, an approved design change may affect the construction documents. A change implemented in the field may change the cost of the negotiated agreement and affect regulatory permits and approvals.

Regulatory or design engineer revisions can affect the design plans. Field inspections may change the construction documents to reflect an engineering approach that can accelerate the schedule or reduce costs.

All changes to the construction documents affect the engineer, government inspections, developer, contractor, and houseline contractors. Any change to the construction documents will affect regulatory permits, approvals, and ultimate purchasers of the lots or houses. Any design changes or changes to the construction documents should be cross-referenced to permits and approvals issued for the project. If no conflicts arise, then the change can be implemented without delay.

Budget versus Actual Cost Report

Once the housing project is underway and the contractor is submitting invoices each month, a budget versus actual cost report should be distributed monthly. A budget versus bid report showing the differences between the budget and the actual bid proposal is outlined in Appendix 18-7. If there are any negative differences between the budget and the bid price, the price of the product being sold may need to be adjusted to meet profit goals.

Summary

Keep in mind that project costs change for a number of reasons. The following is an explanation of some of those reasons.

Unanticipated Site Problems or Conditions

During the site analysis phase, problem areas should be identified for further investigation and analysis. The problem areas are usually not incorporated into the initial conceptual land plan owing to the timing of the purchase of the property. Unanticipated site constraints, such as soil conditions or using GIS topography rather than an actual topography survey, may increase costs.

Unanticipated Permitting or Approval Requirements

Delays in the regulatory process can be expected, but unusual delays are not taken into account in the project schedule and in the project financials. If a plan requires a city council vote and if they carry the agenda item for any reason, the project can be delayed from two weeks to four weeks or more. Delays in the regulatory process cannot be remedied by the developer. Locking in subcontractor costs may not be viable and may be subject to price increase as time elapses.

Accelerated Design or Construction Schedule

Residential land development and home building activities may suffer in quality as developers strive to accelerate schedules. Sufficient time is needed for checking plans and ensuring all elements of the design are finalized before bids, or at least before construction. As regulations increase, the complexities of the construction documents increase, and any change will have a ripple effect. Unclear plans left up to interpretation by subcontractors are an open avenue for higher costs.

Design Changes

Government staff, subcontractors, designers, and owners can change any aspect of the design with or without a consensus. A design change can reduce costs, but most of the time the design change is an enhancement of the project. This, of course, typically results in a cost increase. Design changes are a fact inherent in the nature of the development business. Minimizing design changes should be a goal; to achieve this goal, all parties involved in the project must be forward thinking and creative to keep costs down and maximize return.

Quality of Materials

Land development engineers select the quality of the material used in construction as a reinforcement of local and state requirements. Most engineers base their selection on their experiences and how materials hold up over time. It is reasonable to assume the higher the quality of the material, the higher the cost. There are a few instances when the material may be equal or higher in quality but at a lower cost, for example, ductile iron pipe (DIP) versus polyvinyl chloride (PVC) or reinforced concrete pipe (RCP) versus high-density polyethylene (HDPE).

Specification Changes

During the land development bid process, contractors provide their input on the specifications listed on the design plans or outlined in a separate specification manual. Their changes could reflect cost-saving ideas without jeopardizing the design intent. Local and state governments change specifications in a more structured program and this should be reflected in the design plans. Clients typically seek ways to lower land development costs, and a change in specifications may achieve that goal. However, the design engineer must approve and issue a change order before implementation. In field specification, changes are detrimental to the project cost basis and the schedule.

Material Shortages

As housing enters into full production from any recession level, material shortages can be expected, because manufacturers also slow down production during down turns. Any shortage is created by excess demand of a product that exceeds the supply of that product. There have been shortages in sand, PVC pipe, and lumber affecting pricing, and in some years a shortage of fuel or water. Accelerating a land development schedule may overcome a material shortage or the cost ramifications of that shortage. The quicker the project is completed, the less impact a rise in interest rates will have.

Labor Shortages

The housing cycle affects the labor market and causes labor costs to increase. Even when housing is slowing down and the economy is affected, labor costs rarely readjust, so prices may fluctuate with the cost of labor. The national housing market engine varies from region to region. It is a fact that in the land development industry, labor moves where the jobs are. As technology continues to change every industry, you should expect to see skilled labor shortages in the land development industry. After the housing bubble broke in 2008, many local land development firms and contractors went out of business, relocated, or downsized. Infrastructure owned and maintained by state governments has seen an increase in funding, and the demand for labor and equipment is met by the land development industry.

Union Labor

In right-to-hire states, land development costs will be lower.

Public Bidding Process

There are instances when off-site improvements are a condition of a permit or a regulatory approval. The municipal request for proposals or request for bids procedure not only slows down a project, but costs are increased. This is especially significant when real estate development projects are dominating land development contractors for fast-paced projects and excellent cash flow. It is best for land development companies to negotiate a phased cash payment for off-site improvements, rather than build the off-site improvement.

Issues with the Design, Cost Estimate, Bid Documents, Contractor Quotes

There are a multitude of issues that can change project costs—any design flaw or omission, unclear note or statement, misinterpretation of the plans, citation of the wrong specifications, and so on. The importance of quality design and quality control practices establishes the reputation of the design team, developers, and contractors. Even if problems are discovered, a negotiated settlement working together is the preferred practice rather than seeking a legal solution, which only adds to the costs, lengthens the schedule, and fosters distrust.

Know Your Costs

Tracking costs from project to project is a must. Understanding cost trends is essential for successful projects. Quality, costs, and schedule are the three key ingredients of a successful project.

Appendix 18-1

RESIDENTIAL LAND DEVELOPMENT SPECIFICATIONS OUTLINE

1. DEFINITIONS OF TERMS

1.1	General
1.2	Act of God
1.3	Addendum
1.4	Bidder
1.5	Change Order
1.6	Contract
1.7	Contra at Documents
1.8	Subcontractor
1.9	Directed, Ordered, Approval, etc.
1.10	Engineer
1.11	General Conditions and Specifications
1.12	Inspector
1.13	Laboratory
1.14	Owner
1.15	Performance and Payment Boards
1.16	Plans
1.17	Proposal
1.18	Froposal Guarantee
1.19	Special Conditions
1.20	Specifications
1.21	Supplemental Agreement
1.22	Surety
1.23	Written Notice
1.24	A.S.T.M Designation

2. PLANS, SPECIFICATIONS AND RELATED DATA

2.1	Intent of Plans and Specifications
2.2	Conflict
2.3	Discrepancies in Plans
2.4	Drawing and Specifications at Sub Site
2.5	Dimensions
2.6	Sampling and Testing
2.7	Shop Drawings
2.8	Quality of Equipment and Materials
2.9	Equipment Approval Data

3. ENGINEER, OWNER, SUBCONTRACTOR RELATIONS

3.1	Engineer's Responsibility and Authentity
3.2	Engineer's Decisions

3.3	Suspension of Work
3.4	Extension of Time for Suspension
3.5	Construction Review of Work
3.6	Field Tests and Preliminary Operation
3.7	Examination of Completed Work
3.8	Subcontractor's Superintendent
3.9	Private Property
3.10	Assignment of Contract
3.11	Owner's Right to Correct Deficiencies
3.12	Owner's Right to Terminate Contract and Complete Work
3.13	Subcontractor's Right to Suspend Work
3.14	Rights of Various Interests
3.15	Separate Contracts
3.16	Subcontracts and Purchase Orders
3.17	Work Drawing an Emergency
3.18	Oral Agreements
3.19	Night and Sunday Work
3.20	Hold Harmless Clause
3.21	Unauthorized Work
3.22	Use of Completed Portions of the Work
3.23	Subcontractor's Responsibility for Work

4. MATERIALS AND WORKMANSHIP

4.1	Materials Furnished by the Subcontractor
4.2	Storage of Materials
4.3	Character of Workmen
4.4	Rejected Work and Materials
4.5	Manufacturer's Direction
4.6	Cutting and Patching
4.7	Cleaning Up
4.8	Ownership of Materials
4.9	Guarantee

5. INSURANCE, LEGAL RESPONSIBILITY AND PUBLIC SAFETY

5.1	Insurance
5.2	Indemnity
5.3	Board
5.4	Patents and Royalties
5.5	Permits
5.6	Laws to be Observed
5.7	Protection of Persons and Property
5.8	Crossing Utilities
5.9	Sanitary Provisions
5.10	Warning Signs and Barricades

5.11 Field Offices

6. PROGRESS AND COMPLETION OF WORK

6.1 Notice to Proceed
6.2 Contract Time
6.3 Schedule of Completion
6.4 Coordination of Construction
6.5 Protection of Existing Utilities and Facilities
6.6 Changes in the Work
6.7 Extension of Contract Time
6.8 Liquidated Damages for Failure to Complete Work

7. CONSTRUCTION SPECIFICATIONS

7.1 Surveying
7.2 Clearing
7.3 Soil Erosion
7.4 Excavation
7.5 Storm Drains
7.6 Sanitary Sewer / Water Distribution
7.7 Curbs and Gutters
7.8 Sidewalks
7.9 Asphalt Paving / Roadways and Driveways
7.10 Landscaping

8. MEASUREMENT AND PAYMENT

8.1 Detailed Breakdown of Contract Amount
8.2 Request for Payment
8.3 Engineer's Action on a Request for payment
8.4 Owner's Action on a Request for payment
8.5 Owner's Right to Withhold Payment of a Request for Payment
8.6 Payment for Uncorrected Work
8.7 Payment for Rejected Work and Materials
8.8 Changes in the Work
8.9 Canceled Items of Work
8.10 Payment for Work Suspended by the Owner
8.11 Payment for Work by the Owner
8.12 Payment for Work by the Owner following Termination of Contract
8.13 Payment for Work Suspended by the Subcontractor
8.14 Release of Liens (Interim/ Final)
8.15 Acceptance and Final Payment
8.16 Termination of Subcontractor's Responsibility
8.17 Satisfaction of Warranty Work After Final Payment

9. INFORMATION FOR BIDDERS

9.1	Conditions of Work
9.2	Qualifications of Bidders
9.3	Preparation of Bids
9.4	Bid Security
9.5	Receipt and opening of Bids
9.6	Analysis of Bid Prices
9.7	Right to Accept and Reject Bids
9.8	Acceptance of Bid and Award of Contract
9.9	Security for Faithful Performance
9.10	Time of Completion
9.11	Substitution of Equipment and/or Material

Appendix 18-2

SUMMARY
LAND DEVELOPMENT CONTRACT PROCEDURE

Prebid Conference With Operational Staff

- Review project specifics.
- Outline construction needs, schedule.
- Highlight contractor recommendations.
- Define time tables.

Bid Process

- Schedule prebid meeting with contractors.
- Distribute plans to contractors.
- Schedule contractor meetings as necessary.
- Prepare spreadsheets, analyze schedule of values against cost estimate.
- Resolve discrepancies.
- Ensure contractors understand scope of work.
- Prepare scope of work, spreadsheets, and recommendations.

CONTRACTS

- Complete final negotiations.
- Select contractor.
- Type, process, and execute contract.

PRE-CONSTRUCTION MEETING

- Meet with field representative, design engineer, contractors, utility companies, architect, and government officials prior to construction start.

Appendix 18-3

CONTRACTOR INFORMATION SHEET

NAME____ DATE ______________

BUSINESS ADDRESS______________________________ PHONE _____________

RESIDENCE ADDRESS_____________________________ PHONE _____________

CONTRACTOR LICENSE # _____________ OCCUPATIONAL LICENSE #_______

FEDERAL TAX ID ____________________ NUMBER OF EMPLOYEES _________

TYPE OF BUSINESS: PARTNERSHIP, CORPORATION, SELF-EMPLOYED, OTHER:___

INSURANCE:

	AMOUNT
GENERAL LIABILITY INSURANCE	____________________
AUTO INSURANCE	____________________
WORKMEN'S COMPENSATION	YES________NO________

GENERAL DESCRIPTION OF COMPANY ASSETS / EQUIPMENT:

____

____

____

____

HAVE YOU WORKED FOR ______ BEFORE? YES __________ NO __________

WHEN _______________ WHERE ____________________________________

TYPE OF WORK YOU ARE BIDDING ON ________________________________

____

REFERENCES:

	NAME	ADDRESS	PHONE
BANK #1			
BANK #2			
SUPPLIER #1			
SUPPLIER #2			
CUSTOMER #1			
CUSTOMER #2			

PREVIOUS YEARS SALES VOLUME $________________________

Appendix 18-4

BID ANALYSIS EXAMPLE

		ABC Construction		Able Builders	
	Description	Unit Price	Total	Unit Price	Total
1.	Mobilization		$28,000.00		$37,700.00
2.	Clearing	37.6@$1000	37,600.00	36@$675	24,300.00
3.	Retention Area	43,966@$1.88	82,656.08	43,706@$1.73	75,611.38
4.	Road Grading	21,927@$.76	16,664.52	21,980@$1.55	34,069.00
5.	6" Comp.Sub-Base	21,927@$.50	10,968.50	24,920@$.53	12,873.70
6.	6" Limerock Base	20,500@$5.67	116,235.00	21,980@$6.15	135,177.00
7.	Prime	19,525@$.19	3,709.75	21,980@$.25	5,495.00
8.	1"Type II Asphalt(S-1)	19,525@$1.93	37,083.25	21,980@$1.97	43,399.60
9.	Type "A" Curb	1,743@$4.95	8,627.85	2,430@$4.70	11,421.00
10.	Valley Gutter	10,575@$4.68	49,491.00	10,397@$5.25	54,584.25
11.	Stabilized Soil	1,011@$6.59	6,662.49	3220@$5.40	7,388.00
12.	Street Signs	14@$103.50	1,449.00	16@$75	1,200.00
13.	Stop Signs	6@$92	552.00	16@$90	1,440.00
14.	Lot Prep	182@$693.81	126,273.42	183@$275	50,325.00
15.	Seed,Mulch,Fert.	58,253@$.30	17,475.90	40,000@$.12	4,800.00
16.	Fill Ditch	4901@$.75	3,676.75	move stockpile	-----
	Subtotal -Interior		**$547,719.51**		**$509.684.93**

RECTIFIED BIDS***

	Quantity	Unit Price	Total	Budget Est.
Mobilization			24,000	30,000
Clearing	36.8	838	30,838	52,750
Detention	45,283	1.74	78,792	68,000
Rd Grading	21,954	1.16	25,467	27,000
Subbase	23,066	0.50	11,533	
Base	21,133	5.65	119,401	
Prime	21,133	0.21	4,438	175,540
Type II	21,133	1.98	41,843	
Type A Curb	1,988	4.83	9,602	
Valley	10,608	4.97	52,722	69,600
Stab. Soil	1,216	6.17	7,503	6,000
Street Signs	16	115.00	1,840	2,835
Stop Signs	15	98.00	1,470	
Lot Prep	182	477.00	86,814	57,750
Seed/Mulch	37,988	.21	7,977	15,022
Fill Ditch			1,838	

*** High & Low Bid Cancel & Avg of two Remaining Bids
If all four bid line items, then an average was used; represents cost of project. $506,078 $504,397

Storm Drainage
135,381
141,562
150,190
178,432
178,533

13,129 =2% -------RECTIFIED 641,459
654,587 +50,000 Surveying
\+ 7,500 Bond =$712,087

=16% -- NEGOTIATED BID 551,051

NOTE:

Rectified		Design Additions	
641,459	+	11,995	= $653,454
			$4597
551,051	+	107,000	= $658,051

<u>Fill</u>
47,000
13,000
47,000

Appendix 18-5

LAND DEVELOPMENT CONTRACT PROVISIONS

1. Description and Scope of Work
2. Specifications
3. Plans and Miscellaneous Details
4. Time of Completion
5. Work Schedules
6. Compensation and Payment Schedule
7. Extra Work
8. Inspections
9. Work by Others
10. Existing Utilities
11. Testing
12. Precedence
13. General Conditions
14. Contractor Knowledge of Site
15. Payment and Performance Bond
16. Shop Drawings
17. Subcontractors
18. Construction Staking
19. Special Conditions

Appendix 18-6

Land Development Contract

PROJECT NAME: ______________________________

LOT __________ BLOCK __________

PROJEC T ADDRESS: ______________________________

____

OWNER ADDRESS: ______________________________

____

____

____

CELL PHONE: ____________________

EMAIL: ____________________

The stated price includes all Federal, State and Local Taxes

DESCRIPTION OF WORK

____________________ (the contractor) shall furnish all labor, equipment, material and supervision necessary to complete all LAND DEVELOPMENT ACTIVITIES INCLUDING BUT NOT LIMITED TO: clearing, earthwork, soil erosion, drainage, water, sewer, building pads, curbs, roads, sidewalks, final grading, street trees, etc. and as required to construct the project known as ______________________________.

SPECIFICATIONS

All work shall be performed in accordance with the specifications indicated on the construction documents and required by the Town and/or City, County, State, DOT, Soil Conservation District, Utility Departments, Fire Department, etc.

PLANS AND MISCELLANEOUS DETAILS:

All work shall be performed per plans and details, latest revision dated ____________________ as prepared by: ______________________________.

SHEET #	TITLE/DESCRIPTION
1 of 9	Title Sheet
2 of 9	Grading, Drainage, Landscaping, & Drainage Area Map
3 of 9	Grading, Drainage and landscaping Plan
4 of 9	Road Profiles
5 of 9	Soil Erosion & Sedimentation Control Plan
6 of 9	Water and Sewer
7 of 9	Construction Details
8 of 9	Water and Sewer Details
9 of 9	Erosion Control Details

TIME OF COMPLETION & SCHEDULE:

The contractor shall meet with the owner and City officials for a pre-construction meeting to establish work priorities and expected methods of work production.

Time of Completion: Calendar days ____________. The contractor shall commence work within ten (10) days after the execution of this agreement.

Schedule: The contractor shall perform all construction tasks in a timely manner in order to complete all work within the time of completion. The contractor shall submit a project schedule upon the execution of this agreement.

COMPENSATION:

This contract is a **LUMP SUM CONTRACT** per line item based on the schedule of values for each work item as listed below.

Special provisions noted: ____

PAYMENT:

1. Each monthly payment requisition shall be accompanied by an executed release of lien.
2. Final payment will be made to the contractor within thirty days of satisfactory completion of the project and acceptance of each element of work by the regulatory agencies. Final payment will not be made until all punch list and deficiencies are corrected.

EXTRA WORK:

It is understood and agreed that the contractor shall not perform any additional work under this agreement without authorization by the owner.

INSPECTIONS:

All work performed under this agreement will be inspected by governmental agencies and the contractor must contact each agency and schedule all inspections. Work will also be inspected by the owner.

Should any part of the work, upon inspection, be found to be deficient and thus rejected, the contractor shall correct all deficiencies as soon as possible to ensure the project remains on schedule. The contractor shall request a re-inspection of the work, if appropriate.

WORK BY OTHERS:

The contractor shall coordinate his work with the utility companies and other contractors on site, i.e. curb and paving contractors.

EXISTING UTILITIES:

The contractor shall be responsible for obtaining all existing utility locations to avoid conflicts and/or damage to those facilities. Damage and any costs associated with the repair of existing utilities shall be borne by the contractor.

TESTING:

The contractor shall be responsible for any and all testing required in the acceptance of the infrastructure. The contractor shall provide testing of the subgrade prior to paving by others.

SPECIAL CONDITIONS:

1. The contractor shall abide by all Governmental codes and ordinances required by Federal, State and Local agencies.

2. The contractor shall be duly licensed by all governmental agencies.

3. The cost to obtain all permits to undertake the above stated work shall be paid by the owner.

4. The contractor shall adhere to OSHA regulations, if applicable.

5. The contractor shall provide a certificate of insurance upon the execution of this contract agreement.

6. The contractor shall remove all rubbish, trash, debris accumulated in connection with his work on a continuing basis as work progresses.

7. The contractor agrees that he has thoroughly investigated the site and completed all take offs and associated costs to complete all items specified in this agreement per plan and specification.

8. The contractor shall not sub-contract any work outlined in this agreement without prior approval by the owner.

9. The owner will furnish all construction surveying services for the project. The contractor shall provide detailed drainage information for preparation of as-built drawings.

10. This agreement considers dewatering, select backfill, engineering, and hazardous material removal as extra work items.

11. All changes and/or additions or extras shall be approved by the owner in writing before the commencement of work and the resolution of pricing by the contractor.

12. Other specific conditions of this contract as an example –

a. Silt fence will be paid based on field measurement by the owner.

b. Contractor shall be responsible for payment of restaking if caused by contractor

c. The area from the curb to 5-ft. beyond the road right-of-way shall be graded, top soiled and upon the completion of the curbing.

d. Stone bedding charged at $20.00 per ton in place.

e. Price does not include any excavation below subgrade or placement of suitable materials below subgrade

f. Compaction, if required, shall be charged at $1.60 per LF x VF lift.

g. Soil stabilization, i.e. seeding shall be done by others

13. Contractors bid estimate to be made a part of this agreement and the schedule of values shall be used for invoicing for work in place and payment of such work

AGREED TO BY:

CONTRACTOR

______________________________ Date:____________________

____

____

____

OWER

______________________________ Date:____________________

____

____

____

File Reference: ____________________

Appendix 18-7

LAND DEVELOPMENT
BUDGET VERSUS BID REPORT

Project Name:
Date:
Prepared By:
Reviewed By: Distribution: Field Office

Description	Budget Date: By:	Contract Amount Date: By:	Variance	Contractor
Storm Drainage	$ 86,000.00	$83,270.00	$2,730.00	XYZ Co.
Sanitary Sewer	$ 18,000.00	$19,900.00	($1,900.00)	XYZ Co.
Surveying	$ 10,000.00			
Electric Service	$ 10,000.00			
Road Paving	$ 35,000.00			
Soil Erosion	$ 7,500.00			
Excavation	$ 15,000.00			
Street Signs	$ 1,000.00			
Street Trees	$ 1,000.00			
Curb & Gutter	$ 5,000.00			
Water Distribution	$ 20,000.00			
Clearing	$ 22,000.00			
Sidewalks	$ 1,000.00			
Other	$ 16,000.00			
Subtotal	$247,500.00			
10% Contingency	$ 24,750.00			
TOTAL	**$272,250.00**			

Chapter 19

Land Development Project Management

After the land development contractor has been selected, the builder/developer must now focus on managing the contractor until work is complete. Strive to complete the project on time, within budget constraints, and without field problems. This chapter reviews practical management techniques to oversee field operations (Figure 19-1).

Communication

A daily diary should be maintained by the field superintendent managing the day-to-day activities of the land development contractor. The diary is a historical accounting of the project and substantiates the actions of the land development contractor. The diary should be specific on matters of compliance with the contract agreement. Each day, all land development issues must be recorded and communicated to all involved parties; this includes the municipal inspector and the contractor to ensure that issues can be resolved without delay.

There are several ways to keep track of land development work as it progresses through to completion. The contractor should submit an updated construction schedule with each draw request. Each invoice and schedule should be submitted to the field office—not to the main or corporate office. The draw request and the construction schedule must be reviewed and approved by the personnel responsible for managing that contractor. With each draw request, a simple accounting of the actual costs versus the budget amounts should be analyzed before the invoice is approved and submitted to the corporate accounting office for payment.

Figure 19-1. Field management. Developer personnel or representatives supervise construction activities and approve contractor invoices in accordance with the negotiated contract. Typically, developers will not pay the contractor for material delivered and stored on site. Source: David E. Johnson.

Schedule

Appendix 19-1 shows a land development progress schedule. This land development progress schedule should span the entire infrastructure construction phase. Each listed activity corresponds with the operational requirements of the project. Certain elements of this schedule can enhance the management capabilities of the field staff.

The column heading on the schedule illustrates the time span for each land development activity. As each draw request is submitted, indicate the actual work completed during the past payment period. If a land development contractor with sufficient capabilities is hired, it is not uncommon for more than $100,000 in land development work to be completed in any given month. This level of work varies from contractor to contractor and site to site, but in land development, money can be spent very rapidly. Thus, with each monthly schedule and invoice, the field staff can fully understand the value received from the contractor versus the value requested. Without question, a contractor should not be billing beyond the value received. However, most land development contractors invoice for the work scheduled to be completed during that payment period. It is common practice to retain 10% of the invoice amount each month.

Other blocks within the schedule show work stoppages, delays, activity completions, and a listing of milestones, which are important to review with the progress of the work. As an example, a milestone would be the completion of the first housing pad or the completion of an off-site sewer line, so houses can be hooked up to the sewer. For each work activity and each work period, indicate the percentage of work completed on the schedule. This will clearly show where the work is behind schedule or on schedule.

This schedule is an excellent tool for communicating with a financial institution. Also, field personnel can add important general notes. This master schedule should be updated on a regular basis and copies distributed to staff.

Field Office

The field office should implement another tool of keeping track of work progress. A set of engineering plans should be spliced together showing the complete road layout; lot layout; and water, sewer, and drainage facilities for the project. This overall engineering plan should be tacked to a wall, highly visible to all visitors to the field office. This is especially important for government inspectors, in-house management officers, and outside professionals reviewing and approving construction activities. This field management plan can be updated to show various aspects of the project by the following:

- Color-code all elements of work that are 100% complete, that is, installed and approved by field personnel. As an example, color-code the water lines blue for the portion of the water line installed. Red could highlight the sewer line installed within the project.
- Note the date that the work was completed next to the color-coded activity.
- Include the initials of the field personnel approving the work next to the date.

- Include on this plan the electrical and telephone distribution lines, conduit crossings, and transformer box locations. Color-code these facilities as well to indicate completed work.

This plan also illustrates the work accomplished during a payment period. When an invoice is submitted to the field office, the staff can use the daily diaries and field management plan to understand the extent of the work completed. All invoices can be accurately and appropriately reviewed for accuracy based on field documentation recorded in the construction office.

Invoices and Payment

Keeping track of the work completed by contractors and correlating it with the submitted invoices requires careful monitoring by the builder/developer. Invoices should be accompanied by adequate supporting documentation, and the contract document should clearly explain the responsibilities of the contractor for invoicing and payment.

Schedule of Values Analysis

A schedule of values analysis is provided in Appendix 19-2. This form should be completed by the contractor and submitted with the invoice. This schedule includes the contract amount, any net changes approved, and change orders executed by the contractor and the company. It includes descriptions of work that was negotiated as line items within the contract agreement.

Payment Invoice Summary

The payment invoice summary provides a snapshot of the overall project that can be reviewed and analyzed against the work progress. (A sample invoice is included in Appendix 19-3.) The summary shows previous payments that were approved less the work in place. The sum of the two figures indicates the amount of the work completed. The retainer ensures that warranty issues and quality control of the project are appropriately addressed, given that the contractor should provide a warranty of the work for one to two years.

Release of Lien

A release of lien, as shown in Appendix 19-4, should be attached to every payment request submitted by the contractor. Review the lien laws of your state to ensure that this portion of the contract is enforceable. The release of lien is important, because it is the vehicle by which the contractor notifies the builder/developer that there are no payment problems with any subcontractors or material suppliers. If the release of lien is not submitted, the progress of the project could eventually be affected.

Payment

In addition to the release of lien, a section within the contract document titled "Payment," further identifies the responsibilities of the contractor regarding invoices and payment. Sample language for the contract could read as follows:

- The contractor shall submit to the company every month—unless otherwise specified in the contract or on demand—a written, itemized bill for labor, equipment, and services furnished and materials installed during the preceding month, together with a statement that all indebtedness incurred by the contractor for labor, equipment, materials, and services included in such bills have been paid by the contractor. The company shall have the right to require, as a condition precedent to making any payment hereunder, evidence from the contractor in a form satisfactory to the company that any such indebtedness has been paid and that the contractor has met all obligations with regard to the withholding and payment of taxes, social security payments, unemployment compensation contributions, and similar payroll deductions from the wages of employees.
- The company shall pay such a bill in accordance with the rates, prices, or sums set forth in the various supplements hereto within two weeks after approval of the same by the company, unless otherwise specified in the contract. The contractor must bill the company for extra work over and above that for which payment is specified herein within 30 days after the completion of such work. Otherwise the contractor shall be deemed to have waived the right to payment for the same.
- Notwithstanding the foregoing, it is agreed that any payment by the company of any sums hereunder shall not be deemed to constitute

approval or acceptance of the labor, equipment, materials, or services furnished hereunder.

- Notwithstanding the waiver of mechanics' claims or liens hereinafter referred to, the company shall have the right before making final payment or payments hereunder to require the contractor to deliver to the company a general release of liens in a form satisfactory to the company covering all items of labor, equipment, materials, and services furnished.

  Anything herein contained to the contrary notwithstanding, the company reserves the right to make payment directly to laborers, material suppliers, subcontractors, sub-subcontractors, or any subcontractors or sub-subcontractors of any of them for or on account of work performed or materials furnished under this agreement. If such payments are made in good faith and on reasonable evidence of their validity, the company shall have no liability in connection therewith and shall deduct such payments from any balance owed to the contractor.

Progress Meetings

As each land development project progresses, the field personnel should establish regularly scheduled contractors' meetings. In addition, developers should schedule weekly project meetings on Monday afternoons to analyze the prior week's work and handle any problems that may have arisen over the weekend. The same meeting is used to establish work priorities and issue the new progress schedule for the coming week.

These meetings should be held on a weekly basis regardless of the level of work. Set times, dates, and places in advance. Prepare a meeting agenda, starting with payment problems, work schedules, regulatory issues, change orders, and an open discussion for the benefit of the field personnel. If held on a regular basis, these meetings should last no more than an hour.

Field Extras

Time delays are a problem in residential land development projects, and as a result of delays, additional work may be required of the contractor. Controlling extra fieldwork is critical to the project's financial success. Extras are caused by a difference between the scope of work negotiated and the actual work required that arises from changing field conditions. Field extras are either a result of inexperienced personnel reviewing and negotiating a contract or a result of extenuating circumstances not discovered during the field site analysis work.

Construction extras arise for many reasons:

- Regulatory requirement imposed by a government engineering inspector,
- Design change that enhances the quality of construction,
- Unknown problem not discovered until under construction,
- Problem owing to weather conditions,
- Problem with design, or
- Poor quality of work from subcontractors.

In the budget process, add a contingency line item for extras encountered during construction.

When the need for a field extra arises, first determine *what* the problem is, then investigate *who* may have caused it. Start with the contract documents and determine how the field condition led to a contractual problem. Use the field superintendent's daily accounting of the work progress, weekly reports or diaries, and the project plan. Discuss the problem with the contractors to understand their positions on the causes and solutions. Ask why the problem occurred.

Land development for residential projects is learned through experience, with some of the most important lessons coming from bad experience. Field problems can be minimized during the investigative and design phases of a subdivision project. Remember that land development contractors profit from construction extras, and most contractors have the view that developers do not investigate the site properly and extensively enough to know all the site conditions that should be addressed in contract negotiation. Later, when extra work does occur, contractors have their equipment and labor on the jobsite, and they are in a better position to negotiate for the work.

Some contractors will seek a substantial profit margin for work to overcome a field problem and to keep the project on schedule. For this reason, each problem must be assessed by the field staff, engineer, and surveyor. Along with the contractor, they should recommend a design solution to a problem. This design, if it is outside the initial scope of work, should then be given to the contractor for a price. For each problem, negotiate a lump sum fee for each scope of work; the lump sum contract agreement becomes an addendum to the contract and describes the work to be done and the price for the work. If a developer permits the contractor to proceed without a clear design solution or a negotiated lump sum, the cost of that work usually ends up higher, and the contractor's profit margin will far exceed the contingency budgets established at the outset of the project.

Construction extras, unfortunately, always seem to surface when production is at its peak and the need to keep on schedule is at its greatest. All parties need to proceed and resolve field conditions in an expedited fashion. A contractor's experience in providing a competitive price, mobilizing its equipment and material, handling field problems, and scheduling work consistent with company goals is an important ingredient of a successful project. Analyze the cost of the extras against the budget on a monthly basis. Assessing the ramifications of the cost caused by extras may increase the land development budget and then require an increase in the sales price of the house or lot. If sales price adjustments are needed to accommodate the cost extras in land development operations, this adjustment must be made immediately by the developer and communicated to the sales staff. Price increases should become effective on a set date, making monitoring which homes will be affected easier.

Change Orders

A change order request form, as shown in Appendix 19-5, should be filled out by field personnel and submitted for corporate approval. The form shows the project specifics and the individual proposing the change order. The actual jobsite conditions, reasons for the design change, clear justification of the change, and its cost should be sent to the accounting department for assessment of the budget ramifications. On occasion, the contractor must proceed with correcting a problem without authorization. The remaining portion of the change order form provides for a description of the work to be performed and by which contractor. The change order request is submitted to the field superintendent for review and approval before proceeding with land development activities.

Appendix 19-6 shows a change order addendum to the contract. This one-page contract addendum describes the work to be performed. This change order contract forms a part of the base contract agreement, so all the base contract provisions and requirements are valid.

Appendix 19-1

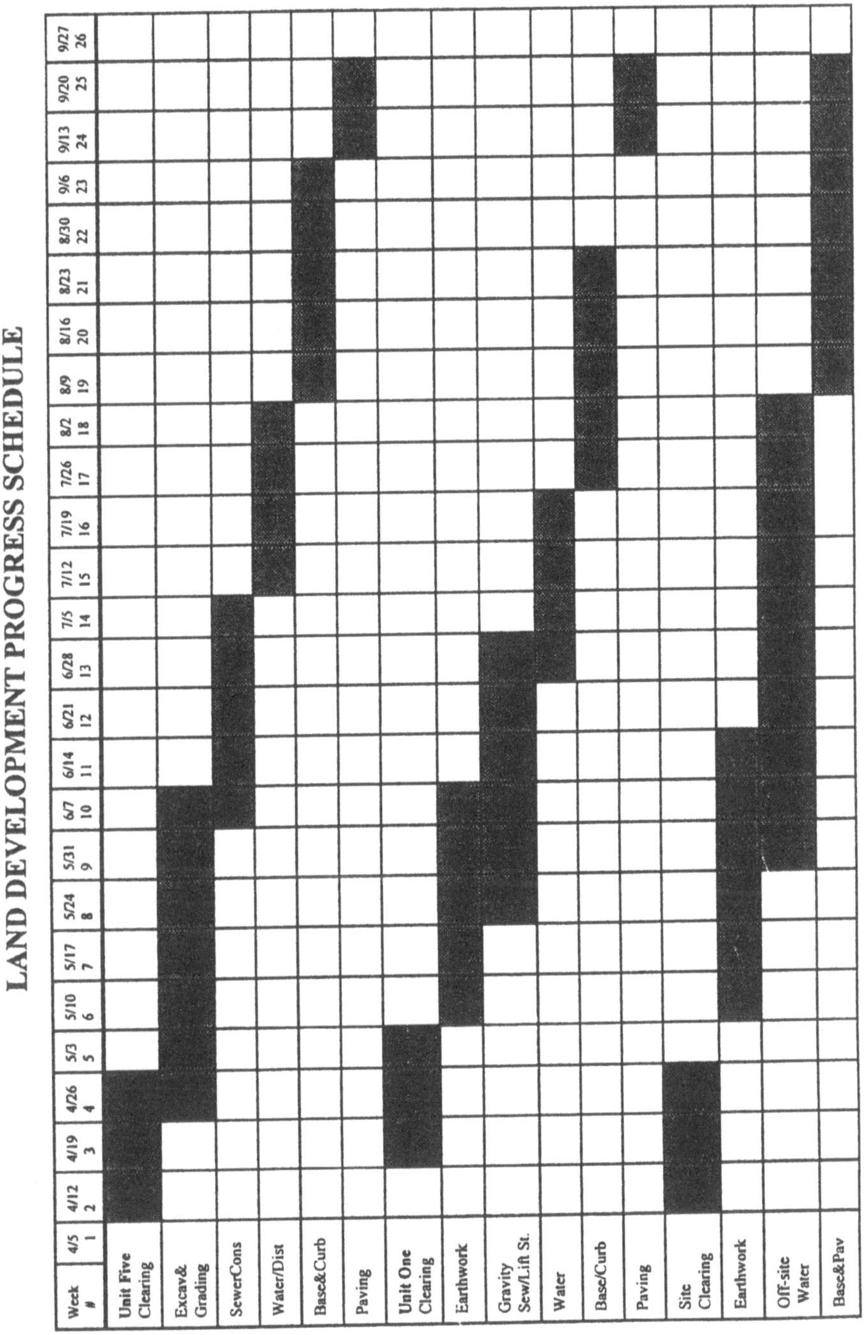

Appendix 19-2

SCHEDULE OF VALUES ANALYSIS

Project :____________________ By:____________________

Location:____________________ Date:____________________

CONTRACT		INVOICE PERIODS			
% Complete	Contract	3/15-5/15	5/15-7/15	7/15-8/15	8/15-9/15
	Addendum 1	$973,682			
15%	Invoice 1	$142,593	$831,089		
22%	Invoice 2		$66,868	$764,221	
24%	Invoice 3		$24,662	$739,559	
	Change Order1		$83,820		
50%	Invoice 1			$41,910	$41,910
				$0	$41,910
	15-Jul Work Order 00908			$2,040	
	Construct Pad Sequence 312			$2,040	$0
% Work Completed	Activity	Invoice 1 3/15-5/15	Invoice 2 5/15-7/15	Invoice 3 7/17-8/15	8/15-9/15
100% $13,986	*Clearing* 12.6 acres @	$10,000	$3,986		
		$3,986	$0		
	Rough Grading				
50%	5,730 cy @ $8595	$3,255	$5,340		
75%		$1,245	$4,095		
50%	*12" Stabil. Subgrade* 9,517 sy @ $16,180			$6,180	$10,000

1 in. = 25 mm; 1 ft = 30 cm; 1 sq yd = 0.8 m^2; 1 acre = 0.4 ha

Note :
1. Prepare a hand log for each contract
2. Monitor schedule of value by contract

Appendix 19-3

PAYMENT INVOICE SUMMARY

TO: ____________________ PROJECT: ____________________

FROM: ____________________

CONTRACT FOR: ____________________ DATE: ____________________

---

1. Original Contract Sum $____________________
2. Total Change Orders $____________________
3. Total Contract Sum to Date $____________________
4. Total Completed $____________________
5. 10% Retainage of Total Completed $____________________

6. Total Earned Less Retainage $____________________
7. Less Previous Payments $____________________

8. Current Payment Due $____________________

9. Balance to Finish, Plus Retainage $____________________

Contractor: ____________________
By: ____________________
Date: ____________________

State of: County of:
Subscribed and sworn to before me this Day of
Notary Public:
My Commission expires:

Received By: ____________________ Date: ____________________
Release of Lien Attached: ____________________

Appendix 19-4

AFFIDAVIT AND RELEASE OF LIEN

Before me the undersigned authority personally appeared. Who after being by me first duly sworn, deposes and says that:

1) He is ________________ of ____________________ doing business in the State of ______________________________, hereinafter called Contractor.

2) Contractor has heretofore furnished or caused to be furnished labor, material and/or services for the construction of certain improvements on the following described real property at

DESCRIPTION	INVOICE #	DATE	AMOUNT

3) Affiant represents that all persons and firms who furnished material, labor and/or services incident to the above have been paid in full except the following: (If no exception, insert the word NONE)

NAME	ADDRESS	AMOUNT
DUE		

4) The undersigned affiant for and in consideration of payment to contractor in the above invoice amount and all other previous payments paid to contractor, does hereby, for and in behalf of the contractor, waive, release, remise and relinquish the contractor's right to claim, demand, or impose a lien or liens for work done or material and/or services furnished or any other class of lien whatsoever, on the above described real property.

5) The affiant herein does hereby represent that he has authority to execute this Release of Lien for and in behalf of the contractor.

6) The affiant herein makes this Affidavit and Release of Lien for the express pupose of inducing ______________________________, a Delaware Corporation, to make payment and disbursement to the contractor in the amount stated above.

7) This Affidavit and Release of Lien is made by affiant with full knowledge of the applicable laws of the State of __________. In addition to such rights as may be afforded under said applicable laws, affiant, individually and as authorized agent of contractor, expressly agrees to indemnify and save ______________ harmless from any and all actual costs and expenses, including reasonable attorney's fees, arising out of claims by laborers, subcontractors or materialmen who might claim that they have not been paid for services or material furnished by or through the contractor on the above-described property.

8) The consideration for this Affidavit and Release of Lien is payment of the above-referenced invoice and is expressly conditioned on receipt of such payment.

SWORN to and subscribed before me
this day of

NOTARY PUBLIC AFFIANT
My commission expires:

Appendix 19-5

INITIATOR CHANGE ORDER REQUEST

Project Title____

Project No.__________ Contract No.____________ Contract Date ____________

Contractor ______________________________ ☐ Owner

Proposed By:______________________________ Date ______ ☐ Engineer

Submitted By:______________________________ Date ______ ☐ Construction

Actual job conditions in area of proposed change:

____

____

Change order justification:

____

____

____

Contractor authorized to proceed with this change ☐ YES ☐ NO on ______________(Date)

Other contracts involved are as follows (List by Contract No.):______Is Dwg Req?☐NO ☐YES________ (SHEET NO.)

Description of Work to be Performed:

____

____

____

____

____

____

____

____

____

____

____

Appendix 19-6

CHANGE ORDER CONTRACT

FORMING PART OF CONTRACT #____________ DATE:________________

EFFECTIVE DATE:_______________ SUBCONTRACT #:________________

CONTRACTOR: XYZ____________ ADDENDUM #: __________________

SUB-CONTRACTOR: _________________ SUB-CONT PHONE:__________

VENDOR # __________________ INDIVIDUAL CONTACT:____________

FOR WORK AT: ____

CATEGORY OF WORK:

1. DESCRIPTION AND SCOPE OF WORK:

2. SPECIFICATIONS

   All work shall be performed in accordance with the following specifications documents:

   A. XYZ General Conditions Date:
   B. Municipal Utility Authority Specifications
   C. State DOT Standard Specifications for Road and Bridge Construction

3. COMPENSATION

SUB-CONTRACTOR:
BY:______________________
TITLE:____________________
DATE:____________________

CONTRACTOR: XYZ CORP
BY:____________________________
TITLE:__________________________
DATE:___________________________
REVIEWED BY:____________________
DATE:________________
PREPARED BY:____________________
DATE:________________

Chapter 20

Land Development Field Operations

Land development field operations must be properly managed to ensure that quality construction is maintained at its highest level (Figure 20-1). A strong field management team results in adherence to a project schedule and enforcement of budget constraints. This chapter presents practical techniques for managing land development field operations.

Construction Surveying

Residential land development activities are undertaken with field guidance provided by surveying crews. Surveyors provide construction stakeouts for the clearing, excavation, street grades, water and sewer control stakes, electrical and phone crossings, storm drainage lines, curbs and gutters, and paving operations. The surveying for housing construction should be a separate contract item. Daily project surveying can be documented by using Appendix 20-1.

The land development contractor, in many states, will not accept the responsibility for construction stakeout work. In some states, however, it is general practice for the contractor to subcontract surveying services. Many contractors prefer to subcontract survey control and retain responsibility for the construction stakeout for residential projects. If the contractor accepts the responsibility for surveying, the owner needs to manage only one firm. Any land development field problem would be discussed with one company, rather than trying to determine which subcontractor is responsible. If the contractor does not accept the responsibility for surveying, the owner can negotiate the surveying services fee at a lower amount, thus saving the profit, overhead, and administrative costs the contractor would charge for managing the surveying company.

The surveying company personnel establishes their own survey control in the field. This survey control is used to set all grade elevations for the infrastructure. If the contractor subcontracts a surveyor, it is advisable to retain a

Figure 20-1. Land development operations. Quality land development construction completed on time and within budget is the primary goal of a developer. Quality construction minimizes government punch list items and expedites the release of performance bonds.
Source: Sidney T. Burns. Due Diligence Associates, Parkland, Florida.

different surveyor to perform the houseline surveying. This provides a double check of the survey control, elevations, and alignment of the infrastructure. If the same surveying company is used for land development and houseline construction, any discrepancy in the field may not be discovered, and the problem will compound as the construction progresses. If a different houseline surveyor company is used, those problems can be discovered with full disclosure and corrective measures made. This check-and-balance approach ensures the layout is consistent with the design plans.

Clearing Operations

Clearing operations can proceed once the surveyor has marked the clearing limits within the street right-of-way, as well as on the lots to be cleared for housing construction. Most surveyors mark the clearing limits by using flags or painting a mark on the trees to be saved at the limit of clearing. The clearing subcontractor works within those clearing limits and proceeds with the stumping and chipping operations in a continuous manner. The clearing operations should not be interrupted for other tasks. As many lots as possible should be cleared at one time for housing operations to maximize the equipment on site.

The land development contract should specify the methods that the contractor will use to remove the stumps from the site. Some contractors use a stump grinder and mix the grindings with the topsoil. Others chip a majority of the trees. It is important to locate the chip pile in an area that will not hinder subsequent land development or housing construction activities.

In addition to the right-of-way and Phase 1 lot clearing, the contractor should also clear a few lots beyond the first phase. The additional area provides a location for the storage of materials, staging of equipment, construction trailers, and administrative trailers on site. It is advisable to reserve one of the better lots for the field offices.

Selective Clearing

The owner's representative should walk the centerline of the right-of-way as well as each lot located within Phase 1 of construction. If a significant number of trees within pad areas or within rights-of-way should be saved, this information should be conveyed to the design engineer and to the regulatory agencies. In many cases, the saving of trees located in building pads and right-of-way areas becomes almost impossible. However, there are several ways to save specimen trees if the species and size of the tree warrant such an action, including improved grading operations, tree wells, and underdrains.

Trees that should be saved can be transplanted to buffer areas or to staging areas for future use. Tree spades come in various sizes to handle significantly sized trees. The marketing potential of saving mature trees within a subdivision cannot be overemphasized, especially when new vegetation purchased from a wholesale landscaper cannot match the maturity of transplanted trees. Many tree-spade operations can be paid for from the savings resulting from the transplanting operations. A mature street scene can be instantaneous and has high market appeal to new home buyers.

Earthwork Operations

Once the site and lots have been cleared, earthwork operations commence. A surveyor, soil scientist, and engineer should become involved in daily earthwork operations. They are the owner's third-party inspectors on the jobsite. The forms presented in Appendixes 20-2 and 20-3 can be used for reporting daily and weekly construction activities. Once the topsoil is stripped from the lots and rights-of-way, it is stockpiled in a cleared area selected by the owner. The right-of-way topsoil is usually stockpiled in an area where it can be accessed by pans or loaders and trucks. The topsoil being stripped from each individual lot is usually stockpiled on the higher portion of the lot to facilitate drainage. Once the topsoil is removed, soil scientists can assess the type of material that exists on the lot. If unsuitable material is located beneath the housing foundation area, the soil scientist determines the extent of the work required to ensure the structural integrity of the foundation. Soil scientists should sign and seal their soils investigation work; they should also inspect the contractor's excavation and removal of all unsuitable material.

Mass filling, cutting, and rough grading of each site is undertaken according to a schedule that should be met without undue delay. Weather will affect earthwork operations and the project schedule. Most delays can be managed properly, and the schedule can be maintained with the cooperation of the earthwork contractor and experienced and knowledgeable operators. The contractor should not deviate from the design plans without an approved change order. When the earthwork contractor's operators can handle the material one time only, everyone benefits. If the developer can show the contractor how to make money on the jobsite, the contractor will approach the project in a positive manner. The net result is profits for the developer and the contractor.

Houseline Land Development

The surveyor locates the house foundation by setting four corner foundation survey stakes with 15 ft offsets and then indicates the cut-or-fill requirements for the pad construction. If a cut for the foundation is specified, the resulting fill material should be used on that specific lot, if possible. Therefore, the earthwork calculations should take into account the fill generated from foundations, driveways, and other flat work on each particular housing lot. The pad should extend 10 ft beyond the perimeter of the house foundation for ease of houseline construction and accessibility for construction of the home. The clearing limits around each home should be a minimum of 15 ft around the house.

After the pad is completed and before the start of foundation work, a soils consultant should take soil compaction tests on the building pad and then certify that the pad is in compliance with the applicable building codes. The rough grading of lots should also be completed with the use of surveying grade stakes. The residential land development contractor should hire equipment operators experienced in housing projects. Building pads should be compacted with rollers to achieve compaction that complies with building code requirements. The soils consultant should also inspect all fill areas to ensure that the operation is in compliance with lift requirements and compaction standards. Once the fill areas have been inspected, tested, and approved by the appropriate regulatory agencies, backfilling within roadway sections and foundations is permitted.

On completion of the final grading, a landscape subcontractor moves the stockpiled topsoil and spreads it on the lot and the right-of-way area. This area is seeded or sodded to complete the lot preparation. The houseline superintendents must have the outside houseline work, such as driveways, patios, and sidewalks, completed before the final grading and seeding or sodding take place. In the tree-save areas, the landscape subcontractor may use the wood chips generated from the clearing operations, thus reducing the need for additional seeded or sodded areas. Residential lots left in a more natural condition will be more appealing and marketable to a majority of buyers.

The balancing of cut-and-fill earthwork operations is essential to the success of the project. Any soil material imported to the project site should be certified for its use as structural material in roadways or foundation areas.

If a substantial number of trees and stumps were cleared from the project site, the grubbing operations will leave significant voids. These voids may require additional fill over and above what was calculated by the engineer. Thus, during the budget process, the fill quantities should be increased by 15% to accommodate this type of clearing operation and compaction of the material. This percent increase is normally determined by the soils consultant and is based on the site soil conditions.

Erosion and Sedimentation Control

Before earthwork operations start, soil erosion and sedimentation control measures need to be implemented in the field. This activity usually includes a stone tracking pad at the access point onto existing roadways. This stone pad is designed to remove any mud, dirt, or debris from the trucks, machinery, or equipment leaving or entering the project site. Installation of a silt fence is another soil conservation measure that protects adjoining properties or wetland areas from sedimentation.

The land development contractor should install inlet protection over the inlet grates as needed. Inlet protection will filter stormwater before it enters the inlets and is subsequently released from the project site. This can be accomplished cost effectively by using a mat with stone over the inlet grates or by using hay bales. The contractor should maintain these protective measures to ensure compliance with soil conservation district requirements. For earthwork operations, the contractor should be required, by contract, to provide street cleaning on a continuous basis to prevent sedimentation from entering the storm drainage system, control dust during dry periods, and ensure the jobsite has an acceptable appearance for any prospective home buyers.

Storm Drainage

The first drainage installations should be the stormwater management facility so it can also function as the soil and erosion sedimentation facility (Figure 20-2). Seed the stormwater basin immediately after it is final-graded; this minimizes the future maintenance requirements of the facility. Some municipalities require the fencing of stormwater management ponds and access gates and roads for maintenance purposes. Choose the proper fencing material if it is critical to the overall appearance of the project. Depending on the location of the stormwater management pond, a typical chain-link fence may be appropriate.

However, if the facility is located in and among housing units, a black or green vinyl chain-link fence is a better choice, as the color blends into the background. This type of fencing and evergreen landscaping transforms the basin into a positive attribute of the project.

Figure 20-2. Drainage facilities. Erosion control and stormwater management facilities must be installed and operate properly to adequately control runoff during the construction process. Erosion and sediment are visible and indicative of the lack of quality control. Source: David E. Johnson.

Utility Installations

The soils engineer should use the soils testing information to determine the seasonal highwater table and the necessity for well pointing during underground utility installation. Dewatering is an expensive operation and needs to be addressed in the contract before commencement of the work.

Conflicts between the sewer line and other utilities are found at crossings and when parallel lines are less than 10 ft apart. Review with the contractor all potential conflict areas illustrated on the plans. Make sure the utility crossings, horizontally and/or vertically, are not in conflict before construction. Many sewer and drainage lines are set at slopes governed by outlet elevations, and the locations are difficult to change in the field. Waterline elevations and locations are much easier to change in the field if a conflict occurs.

Electric, Phone, Gas, and Cable

The electric, phone, gas, and cable utility companies use their own crews or subcontractors for installations. The difference in company crews versus subcontractors can lead to substantial differences in time, quality, and effort in residential community installations. Utility companies are often unilateral in their design approach to a project. Their plans are specific enough that any conflicts that may occur in the field should be readily apparent from their design plans.

Conflicts in the locations of streetlights, transformers, and pedestals should be checked against the plot plans for each lot. The service locations from a transformer to a lot should dictate the right- or left-hand garage type. If it is customary to install the electrical panel in the garage, the garage should be located on the side with the electrical service and according to the electrical layout design. Otherwise, a charge could be assessed against the lot for additional service length to the house.

Most conflicts caused by electric, phone, gas, and cable companies pertain to the location of their pedestals or transformers within the right-of-way area. Their equipment locations could affect drainage swales, driveways, sidewalks, street trees, grade elevations, and other areas.

Water and Sewer

Once utility services are installed, many water and sewer companies require an underground utility contractor to etch a "W" or "S" into the curb face so the lateral can be located once the project has been completed. Field representatives or the project surveyor should field-locate the valve boxes, clean-outs, and fire hydrants and note them on the project map for future reference. Planners will suggest that clean-outs and valve boxes be located outside of a driveway. This may have been accommodated in the design, but the implementation in the field must be managed properly.

To expedite the utility company's installation of its facilities, contract separately with the land development contractor for the installation of utility conduit for road crossings. Include the cost of conduit road crossings in the budget. This expenditure helps to keep the project on schedule, because the project does not have to wait for the utility company. As long as the conduit meets specifications and is in accordance with regulations, utility companies can install their lines according to their schedule. This is not a problem, as long as the first home has service when needed.

Curb and Gutter Installations

The surveyor provides a field alignment and grade stakes for curb installation. If the project is preplanned with right- or left-hand garage and driveway locations, the surveyor should mark in the field where the driveway is located. Thus, the curb subcontractor can then install depressed curbs, driveway aprons, and sidewalks consistent with the overall plan of the project. The soils engineer should take random compaction tests under the curb sections to ensure the sub-base has sufficient structural integrity to support the curb without deflection or failure. The cost of curb replacement and concrete removal can be substantial.

Communicate with the surveyor on which lots are designated with one- or two-car driveways so that the correct width of the curb cut can be installed. To facilitate houseline construction, include a requirement in the land development agreement that the contractor must *maintain access provisions* to each house. Also, include a provision concerning the protection of the curb from damage caused by large equipment or heavy supply trucks. The infrastructure is extremely vulnerable to damage from suppliers of heavy housing construction materials, such as concrete, lumber, and drywall. Damage to concrete curbs by chipping, cracking, or settling because of the lack of protection will double the cost of curb installation.

Each individual lot needs to be rough-graded to ensure access for houseline subcontractors, material suppliers, well drilling equipment, and landscapers. If proper care is taken and the subcontractors are notified of areas that can be used for access, the potential for damaging the infrastructure is minimized. Proper precautions taken during construction will eliminate conflict or arbitration as a result of infrastructure failure. If proper precautions are not taken, the cause of the infrastructure failure must be determined and actions taken to eliminate the cause.

Pavement

The final pavement should be installed when 90% to 95% of the homes have been closed or are beyond the sheetrock stage. Install street trees on the lot before closing the house. This eliminates the need for subsequent landscaping after it is closed. This also facilitates approval for maintenance bond status.

Once the final paving is completed, request that the surveyor prepares an as-built plan from right-of-way line to right-of-way line. This as-built plan, along with the completion of government punch lists, will expedite the process of having the project released to maintenance bond status. Take pavement depth measurements during the inspection of pavement construction. Use the roadway construction report, as shown in Appendix 20-4, to record the inspection observations. This also facilitates the acceptance of the right-of-way by a municipality or homeowners association (HOA).

Land Development Punch Lists

Completing infrastructure punch list items is a time-consuming, difficult, and costly aspect of land development. The contractors on the project can complete most punch list items with all due diligence. However, a punch list that lingers and is not completed provides opportunity for additional punch list items to be added at a future date. Once subcontractors leave the jobsite, requesting that they return to repair problems on the punch list is difficult. It is better to complete the punch list without delay, so the performance bond can be released. The longer the delay, the more difficult it is to achieve maintenance bond status.

Any street traffic control devices, stop signs, street signs, yield signs, and so on, should be installed in accordance with traffic control standards. If necessary, install speed limit signs for the subdivision. If the residential street design is appropriate, speed limit signs will not be needed because the road alignment, horizontally and vertically, controls vehicle speed within the subdivision.

Appendix 20-1

LAND DEVELOPMENT
DAILY SURVEY REPORT

CONSTRUCTION STAKING:

LOCATION:____

*CIRCLE ONE: Sewer Water Drainage Roadway Curbing Other*______

HOURS CREW ON-SITE FROM ______________ TO ______________

TOTAL HOURS ______________

STATION ____________________ TO ________________________

MONUMENTS: AMT. COMPLETE ____________

LOCATION ____

CIRCLE ONE: Sewer Water Drainage Roadway Curbing Misc. ______

HOURS CREW ON SITE FROM ______________ TO ______________

TOTAL HOURS ______________

OTHER:

LOCATION: ____

REASON: ____

HOURS CREW ON SITE FROM ______________ TO ______________

TOTAL HOURS ______________

Appendix 20-2

LAND DEVELOPMENT DAILY CONSTRUCTION REPORT

PROJECT ______________________ DATE ____________

COMMUNITY ______________________ DAY: S M T W TH F S

CONTRACTOR ______________________ WEATHER: CLEAR OVERCAST RAIN

COMPANY REP: ______________________ TEMP 0-32 32-50 50-70 70-85 85+

AVERAGE FIELD FORCE / EQUIPMENT IN USE OF IDLED

QUANTITIES

LOCATION SEWER	MFG/GLASS	PIPE SIZE	AMT. PLACED	LATERALS LOT BLOCK	MANHOLE #	MISC

LOCATION DRAINAGE	TYPE	PIPE SIZE	AMT PLACED	INLET #	MANHOLE #	MISC

LOCATION WATER	PIPE SIZE	AMT PLACED	LATERALS LOT BLOCK	VALVES/TEES/PLUGS	FIRE HYDRANT

SUMMARY OF CONSTRUCTION ACTIVITIES

FIELD PROBLEMS (WHICH COULD RESULT IN DELAY OF CLAIM)

FOLLOW UP INSPECTIONS OF PREVIOUSLY REPORTED DEFICIENCIES

Appendix 20-3

INSPECTOR'S WEEKLY REPORT

COMPLETED BY:____

PROJECT:____________________________ DATE____________

CONTRACTOR:____

SUBCONTRACTOR: ______________________________________

PROGRESS STATUS:

COMPLETE % __________ CONTRACTOR COMPLETE % __________

INCOMPLETE _______________ INCOMPLETE _______________

MATERIALS SUPPLIED THIS WEEK:____________________________

PREVIOUSLY SUPPLIED:__________________________________

TOTAL SUPPLIED TO DATE: _______________________________

REMARKS:____

Appendix 20-4

LAND DEVELOPMENT
ROADWAY CONSTRUCTION REPORT

PROJECT____________________ DATE __________

COMMUNITY ____________________ DAY: S M T W TH F S

CONTRACTOR____________________ WEATHER: CLEAR RAIN OVERCAST

CO. REP: ____________________ TEMP: 0-32 32-50 50-70 78-85 85+

STREET NAME ____________________ STATION ________ TO

	DEPTH CHECK	WIDTH CHECK	CBR	DENSITY
SUB-GRADE: (CIRCLE ONE) STABILIZED COMPACTED				
SUB-BASE: (CIRCLE ONE) ROAD GRAVEL QP S/C OTHER				
BASE: STABILIZED BASE - TYPE __________				
ASPHALT: FABC TYPE __________				

STREET ____________________ STATION ________ TO ________

SUB-GRADE: (CIRCLE ONE) STABILIZED COMPACTED				
SUB-BASE: (CIRCLE ONE) ROAD GRAVEL QP S/C OTHER				
BASE: STABILIZED BASE - TYPE __________				
ASPHALT: FABC TYPE __________				

MISCELLANEOUS

LOCATION:____

PSI: ____________________ SLUMP:____________________

TAKEN #: ____________________ BATCH #: ____________________

(CIRCLE ONE:) Curb Inlet Flume Other:__________

Chapter 21

The Land Development Business

In the land development business, many people outside the construction trades are associated with designing, permitting, and approving subdivision projects. Land development companies rely on professionals who enjoy the challenges of converting raw land to finished residential lots. The influence of these stakeholders on the success of a land development project is extraordinary. External pressures exerted by various interested parties or stakeholders have a direct bearing on a land development company's ability to earn a profit. Land developers take the most risk but also can achieve the highest rate of return on their investment. Design professionals are involved from the outset to minimize the developer's risk, understand the costs, and implement the project vision consistent with government oversight.

The first part of this chapter surveys these stakeholders and their influence on the land development process. Not only do these groups directly affect a company's financial health, but they also influence the ability of buyers to own homes. The second part of this chapter briefly examines business issues that affect the housing industry.

Stakeholders

Government

Regional, state, and federal governments are structured to oversee projects and to ensure compliance with their regulations, standards, and criteria. Local government, however, controls the growth patterns of its community. The attitudes of local government toward housing vary widely across the United States and even from street to street. Practical and progressive local governments are usually advocates of new housing projects; their positive actions result in communities being developed in a way that benefits those who live and work there.

Communities that limit growth through government restrictions subject themselves to unpredictable economic expansion. Restrictive controls on growth or excessive regulations on development adversely affect only people living in that community. One type of growth control commonly instituted by local governments involves down zoning property. In effect, this action reduces the number of housing units allowed per acre and therefore reduces the supply of buildable land in the community. The law of supply and demand—specifically, the lack of supply—ultimately increases the cost of housing. Moreover, raw land rarely loses value unless that value is affected by government action. In addition, some communities limit growth by passing the responsibility for improving existing community infrastructure to the developers of new housing projects (Figure 21-1). Reducing the availability of land, increasing the cost of development, and inflating the value of land affect the ability of future generations to own a home.

A balanced community encourages business and industrial expansion and prepares for legitimate and practical housing growth. Local government should balance the economic stability of the community for the betterment of society. The strong traditions of living and working in a community, for the community, must change government's grip on housing for housing to succeed as a basic industry of the country.

Figure 21-1. Residential infrastructure. Subdivision improvements are taken for granted until a component fails. Minimizing infrastructure problems starts with quality design, construction, and inspection. Source: David E. Johnson.

Business Community

The business community is active in the Chamber of Commerce and various civic organizations but rarely publicly supports new housing projects. This group of stakeholders should become more involved in the growth pattern of a community. The business community benefits from an active pro-growth government, and business leadership should encourage pro-growth political agendas.

Citizens

Too often, citizens of a community only participate at public hearings by presenting negative points of view toward housing projects. Rarely will anyone who is in favor of a project attend a planning board meeting. It is also common that citizens use emotional issues to persuade planning board members and politicians to reject housing projects. A definite no-growth attitude is present in many communities. Citizens use the "problems" of traffic, lack of or pollution of water, wastewater, environment, safety, security, and so on against new housing development. Even sincere and concerted efforts by the housing industry and others may not be enough to overcome the public no-growth attitude. Thus, the refrain "Not in my back yard!" must be countered by continuing to ask, "Where will my children live?"

Environmental Community

Environmental regulations are the most emotional and restrictive limits on housing developments today and are likely to be so for many years ahead. Environmental issues, including endangered species, endangered plants, wetland protection, floodplains, coastal area protection, and open-space protection, have the greatest effect on housing affordability. Because these regulations are likely to become more restrictive and widespread, builder/developers must learn to work within the environmental community and the regulations imposed at the local, state, or federal levels. Housing development is a positive environmental contribution to society. The builder whose business becomes environmentally "green" will be consistent with the demands of the future homeowner.

Legal Community

The legal structure surrounding the development, building, and sale of housing has become an unbelievable bureaucratic web that entangles progress of the industry. Legal challenges by special interest groups affect the way projects are built and homes are sold. Building companies face liabilities that dampen enthusiasm for building new homes. Builder/developers should seek to negotiate resolution of problems and keep all conflict issues out of the legal arena as much as possible. Doing so saves money, time, and effort so that all energy can be channeled into building subdivisions on time and within budget. A negotiated settlement of a problem remains the best alternative to a legal solution.

Political Community

The political community moves with the tide of public opinion. Successful developers design projects by conforming to all government regulations, reducing the number of variances, and eliminating design waivers. Any project in compliance with the master plan, zoning regulations, subdivision ordinances, and technical standards should be approved regardless of local opposition or political attitudes. Builder/developers will increase their chances of success at the planning board by providing quality living environments for future home buyers.

Homeowner Associations

Once a project is turned over to a homeowners association (HOA), homeowner interests are protected by this central organization, which is formed with financial stability. HOAs tend to seek builder remedies for issues governed by local ordinances, thus becoming adversarial to the developer. In many single-family subdivisions, HOAs are created to own and maintain stormwater facilities because the municipality will not accept the responsibility. However, because HOAs are known to at times overextend their powers, many builders are including common areas in individual lots to eliminate the need for an HOA.

Business Issues

Profits

The profit margin of a development company continues to be restricted by regulations. Land price, value, and cost determine the profit margins obtained on any one particular project. The housing style, products, and project attributes increase or decrease the absorption and affect the cash flow needs of a company. However, land price and value ultimately determine the profitability of the company.

Reputation and Corporate Image

Builder/developers must continue to improve their public image through success. Reputation is supported by achieving quality construction, meeting public commitments, addressing warranty issues, and providing service to their customers. A sense of cooperation with government reinforces a company's reputation within the community. One highly successful marketing technique is to incorporate testimonials of previous buyers, because referrals are the basic foundation for housing sales. An impeccable reputation built by satisfying customers reflects a corporate image of professionalism and leadership within the industry.

Effect on Society

Every industry has a sector that is self-centered, and in the housing industry, this group's lack of sensitivity to the home-buying public unfairly casts a shadow on the building community as a whole. An attitude of distrust, bolstered by a few builders taking advantage of the public, is a difficult roadblock for the building community to overcome. The building community receives few accolades, regardless of the positive effect housing has on society. It is a silent reward for the building community to know that their project provides affordable, safe, and efficient housing for all people.

Design Professionals

Design firms actively involved in the building industry must uphold the same ideals as the builder/developer. A quality living environment is the greatest achievement of any technical designer. With all of the troubles

Figure 21-2. The ease of construction is reflective of the design quality. Implementation according to the design expedites the ability of the developer to achieve maintenance bond status and thus reduce cash flow requirements.
Source: David E. Johnson.

and complexities of the building industry, the completion of a well-conceived project overshadows the history of problems and issues confronted by each project (Figure 21-2).

Land Development Industry Changes

Predicting how the land development industry will change in the future is difficult. Each project is influenced by many people and by excessive regulations imposed at many levels of government. The land development industry will be very slow to change because changes in housing must be well received by all stakeholders before change is realized. Most likely, large regional builder/developers will not be able to compete consistently at the local level. Builders of 10 to 20 homes per year will continue to dominate fringe markets. Developing land for use by builders is not an easy business. The need for business segmentation is not a result of regulations, but it does evolve from regulation.

If you are new to this industry, welcome aboard! The complexities, challenges, and efforts are all worthwhile when a quality subdivision can be constructed and houses built for families who need homes. By implementing the techniques outlined in this book, you, the housing professional, can build and sustain a quality of life that should be cherished by all.

Chapter 22

Land Development Engineering

Most civil and environmental engineering programs at college or university levels are centered on the four primary and historical civil engineering disciplines: transportation engineering, structural engineering, environmental engineering, and geotechnical engineering. There is a multitude of subdisciplines students may select, but land development engineering is typically not available. The level of engineering needed each year to design, construct, and maintain infrastructure to serve 1,000,000 new homes would be

- 1,000,000 new housing starts *per year*;
- 500,000 pairs of homes (assume homes on both sides of the street);
- 40,000,000 linear feet of roadways with associated water, sewer, and drainage facilities (assume an average 80 ft wide lot);
- $400 per foot of roadway for infrastructure; and
- More than $16 billion in construction costs spent each year for residential subdivisions alone and not for projects of nonresidential uses.

Land development engineering is a significant consulting business opportunity that thrives locally with community roots. There will be about 120 million more people born in the United States over the next 40 years, and land development engineers must consider the following questions:

- Where and how will they live?
- Where and how will they play?
- Where and how will they shop?
- Where and how will they work?
- How will they think?

Knowing the extent of the construction costs associated with real estate development, the land development engineer has a significant role in the civil engineering profession. Today, most land development engineers, regardless of career path, are trained on the job, which includes working for a

- Consulting firm that has public and private clients;
- Local, county, state, or federal government agency;
- Quasi-government organization;
- Builder/developer;
- Individual (only one client is needed to start a consulting business);
- Multidisciplined consulting firm providing career path opportunities within the same firm, such as geotechnical, surveying, planning, environmental, civil engineering, construction inspections, material testing; or
- College or university educating future land development engineers.

Civil engineers who have gravitated to or found a purpose in land development engineering rarely change their professional direction once acclimated to this discipline, because most residential projects are designed, constructed, and closed out in a reasonable amount of time.

- Every project is different.
- Every parcel of land is different in shape.
- Every property has different site characteristics.
- All properties have different land development or design constraints.
- The location of every unique property brings different challenges.
- Every project is different, because government regulations change from town to town.
- A client's intended use or zoning criteria changes for each unique project.
- Each client's vision varies from the perspective of a small local builder/developer to the national companies.

Why would you, as a civil engineer, consider land development engineering as your career path? Some reasons are that you

- Interface with a wide range of people of various backgrounds,
- Use all your civil engineering design skill sets on each project,
- See the results of your design in a short time,
- Can influence public policy,
- Can promote sustainable developments,
- Directly affect the way people live in their communities,
- Are an economic force in your community,
- Are a social force in your community,

- Enjoy the engineering complexities of each project, and
- Appreciate the silent reward of a successful project.

Land development engineers also must work with a variety of government agencies at the local, state, and federal levels, including but certainly not limited to

- Planning boards and zoning boards,
- Planning and engineering departments,
- Fire departments,
- Police departments,
- Environmental commissions,
- Shade tree commissions,
- Town foresters,
- Water and sewer departments,
- Building departments,
- Regional regulatory agencies (e.g., soil conservation district),
- State agencies (e.g., DOT or EPA), and
- Federal agencies (e.g., USACE or US Fish and Wildlife Service).

Land development engineers must interface with vested interest groups and other land development professionals on each project. The interpersonal skills of a land development engineer are tested on each project with

- Builders and/or developers;
- Contractors, vendors, suppliers;
- Attorneys;
- Neighbors;
- Members of the public including NIMBY ("not in my back yard") and CARE ("citizens against really everything");
- Land planners;
- Landscape architects;
- Marketing/public relation firms;
- Environmentalists;
- Architects;
- Economists;
- Traffic engineers;
- Geotechnical engineers;
- Surveyors;
- Water and wastewater engineers;
- Realtors; and/or

- Structural engineers.

Primary responsibilities of a land development engineer are to

- Prepare due diligence reports used for the purchase of land,
- Create land plans for residential and nonresidential projects,
- Prepare construction documents,
- Develop cost-effective engineering designs and solutions,
- Influence the profitability of the project,
- Affect the way people live,
- Establish a level of importance for land development projects within the firm, and
- Act as the thread among all design team members.

Graduating civil engineers have an excellent foundation to start work in any discipline except perhaps in the land development industry. Civil engineers need to learn and perfect their capabilities on the job. In 2006, David Young, Ph.D., P.E., Civil and Environmental Engineering Department chair at the University of North Carolina Charlotte (UNCC), recognized that a majority of the graduates were being hired by land development engineering firms. Dr. Young was instrumental in introducing and supporting land development engineering courses as an option for UNCC civil engineering students.

Then in 2017 land development engineering became a concentration within the civil and environmental engineering program at UNCC (Figure 22-1).

According to UNCC's Department of Civil and Environmental Engineering, the civil engineering program is accredited by the Engineering Accreditation Commission of the Accreditation Board for Engineering and Technology (ABET). There are other universities that have real estate courses focusing on land development topics, but UNCC is the only 4-year university to offer land development engineering as an alternative engineering discipline and a concentration within a civil and environmental engineering department.

UNCC's civil and environmental engineering degree programs include

- Bachelor of Science in Civil Engineering
- Master of Science in Civil Engineering
- Master of Science in Engineering
- Doctor of Philosophy in Infrastructure and Environmental Systems

Figure 22-1. The University of North Carolina Charlotte is a public research university located approximately eight miles from uptown Charlotte. UNCC has an enrollment of more than 29,000 students. Source: UNCC website (wwwuncc.edu).

Undergraduate students in the program may also select an optional concentration in the following technical areas:

- Energy infrastructure,
- Environmental/water resources engineering,
- Geotechnical engineering,
- Land development engineering,
- Structures, or
- Transportation.

Course Requirements

Students must take these two courses:

CEGR 3231: Land Development Engineering Fundamentals
CEGR 4185: Geometric Design of Highways

Students must choose two of the following:

CEGR 3235: Land Development Engineering—Advanced Site Analysis
CEGR 3233: Land Development Engineering Studio
CEGR 4247: Sustainability
CEGR 4147: Stormwater Management

UNCC's program will lead the way for other civil engineering departments to consider land development engineering as a viable career path for students.

The following course topics are intended to provide a framework in which discussions may be started with other colleges and universities.

Fundamentals

This course focuses on the basics of the real estate industry and the role of the land development engineer:

- Real estate industry overview
- Panel discussion on land development engineering (panel consists of local professional engineers)
- Four uses of land
- Residential and nonresidential land uses
- Sustainable engineering
- Government policies, regulations, and standards
- Role playing from stakeholder, applicant, and government perspectives
- Site analysis
- Base map review
- Topography analysis and drainage areas
- Art of contouring
- Stockpile grading
- Basin grading
- Drainage swales
- Roads
- Residential lots
- Infrastructure planning and engineering concepts
    - Drainage systems, SWM, and BMPs
    - Roadways for real estate developments
    - Drainage design
    - Water (conflict analysis) and sewer (design and invert chart)
- Value engineering and cost estimating.

Advanced Site Analysis

Students are given a single-family subdivision plan with existing topography and base map information necessary for a complete infrastructure design using all available software, including AutoCad.

- Zoning requirements
- Site analysis
  - Environmental, wetlands
  - Soils characteristics
  - Water and sewer
  - Jurisdictional streams
- Topography analysis
  - Slopes and soils
  - Before development
  - Drainage patterns
- BMP and SWM locations
- Roadway design
  - Access location study
  - Horizontal alignment
  - Typical sections
  - Pavement criteria
- Roadway design
  - Vertical alignment
  - Level terrain design
- Preliminary road grading
- Water line design
- Sewer line design
- Storm drainage design
- Post-development drainage area plan
- Infrastructure conflict assessment
- BMP and SWM facility design
- Basin grading design
- Frontage road improvements
- Site development details
- Erosion control plan
- Clearing analysis
- House location planning and lot fit matrix
- Lot grading
- Quantity takeoff and costs

Studio

This course is unique to most engineering programs because it provides a means by which a design style can be discovered by students while understanding the nuances of design, functionality, and sustainability. A base map and project requirements are provided, and each project is designed and presented by the students. A student's portfolio of project designs is made available to hiring managers. In-class design projects include fire station site plan, commercial strip center site plan, major subdivision layout, and a land-use bubble plan.

- Single-family subdivision
  - Project vision and goals
  - Project site characteristics and constraints
  - Bulk requirements and design program
  - Engineering standards and road design
  - Plan and presentation guidelines
  - Single-family subdivision design concepts
  - Review of final plat and base map
  - Water, sewer, and drainage layouts
  - Land development costs
  - Quantity takeoff, cost estimate, and pro forma
  - Single-family subdivision project presentation
- Multifamily site plan
  - Project vision and goals
  - Project site characteristics and constraints
  - Bulk requirements and design program
  - Engineering standards and road design
  - Plan and presentation guidelines
  - Recycling, garages, fire access, hardscape
  - Budgets, off-site impacts, common area amenities, phasing, master developer
  - Multifamily site plan presentation in a planning board environment
- Vertical mixed-use site plan
  - Project vision and goals
  - Project site characteristics and constraints
  - Bulk requirements and design program
  - Engineering standards and road design
  - Plan guidelines

- Vertical mixed-use design concepts
- Off-site impacts, common areas, phasing
- Sustainable development
- Vertical mixed-use site plan presentation

Storm Drainage Design

A hands-on step-by-step storm drainage design course is applied to a residential subdivision layout:

- Drainage design overview, rational method, TR55, and Soil Conservation Service
- Predevelopment drainage areas and time of concentration
- Post-development drainage areas and time of concentration for single-family projects
- Post-development drainage areas and time of concentration for multifamily and commercial
- Rainfall intensity and soil characteristics
- Runoff coefficient and weighted runoff coefficient
- Runoff discharge calculations using the rational method for storm drainage design
- Ditch section road design and cross culverts
- Inlet locations for different project applications, inlet types, and headwalls
- Gutter flow and allowable roadway spread calculations
- Parking lot design configurations and drainage considerations for commercial and multifamily projects
- TR55 runoff curve number and runoff calculations
- Rainfall distribution, storm events, and regulatory standards
- Peak discharge using TR55 methodology
- Storm drainage design project: single-family subdivision
- Storm drainage design project: commercial project (reference multifamily)
- Soil erosion and sedimentation control overview, standards, and criteria
- Phase I and Phase II soil erosion control design approaches and considerations
- Riprap design and stability calculations
- SWM pond contour grading
- Roadway with ditch section and curb section contour grading

- Parking lot contour grading
- Cost estimates

A more detailed syllabus of each course is available on request. Please give this information to your civil engineering department as a practical opportunity for educating students for the reality of servicing the land development industry. There are other course ideas to consider, such as costs, contracts, negotiation, and public sector dispute resolution. By giving students a firm foundation in land development engineering, public and private employers will recognize a shorter learning curve and an advantage in hiring a land development trained engineer.

Index

About the Author

David E. Johnson, P.E., F.ASCE, has more than 48 years of experience in the land development industry. He is a Fellow and Life Member of ASCE. He holds a Bachelor of Science degree in civil engineering from West Virginia University Institute of Technology. He is a registered professional engineer in Maryland (retired), New Jersey (retired), Virginia, and North Carolina. His experience includes 5 years as a consulting engineer, 12 years as a sole proprietor in land development engineering, 17 years in leadership roles with builder/developers in New Jersey and Florida, 3 years with the National Association of Home Builders (NAHB) in Washington, DC, as manager of the Technical Services Department, and 11 years working in local government public works departments.

Johnson has been responsible for the design, permitting, development, and construction of residential, active adult, assisted living, commercial, professional office, industrial, and master planned projects located in New Jersey, North Carolina, and Florida.

He was a part-time instructor of land development engineering courses in the Department of Civil and Environmental Engineering, The University of North Carolina, Charlotte for 11 years. For more than 35 years, he has promoted land development engineering as a unique discipline within university civil engineering programs. In addition, Johnson has been an ASCE seminar instructor for more than 11 years and has written articles for ASCE and public works magazines, as well as been a presenter of land development and construction topics at numerous workshops held throughout the United States by ASCE and NAHB. He also authored the book, *Fundamentals of Land Development: A Real-World Guide to Profitable Large-Scale Development*, published in 2008.